The Intergenerational Transfer of Cognitive Skills

Volume II: Theory and Research in Cognitive Science

COGNITION AND LITERACY
Series Editor: Judith Orasanu
U.S. Army Research Institute

Becoming Literate in English as a Second Language, Susan Goldman and Henry Trueba (eds.)

The Intergenerational Transfer of Cognitive Skills, Volume I: Programs, Policy and Research Issues, Thomas G. Sticht, Micheal J. Beeler, and Barbara A. McDonald

The Intergenerational Transfer of Cognitive Skills, Volume II: Theory and Research in Cognitive Science, Thomas G. Sticht, Micheal J. Beeler, and Barbara A. McDonald

Metacognition and Reading Comprehension, Ruth Garner

In preparation

Literacy Development: Comparative Studies in the Acquisition and Practice of Literacy, Stephen Reder and Karen Reid Green

The Intergenerational Transfer of Cognitive Skills

Volume II: Theory and Research in Cognitive Science

edited by

Thomas G. Sticht
Micheal J. Beeler
Barbara A. McDonald

ABLEX PUBLISHING CORPORATION
NORWOOD, NEW JERSEY

Printed in the United States of America.

Library of Congress Cataloging-in-Publication Data

The Intergenerational transfer of cognitive skills / edited by Thomas G. Sticht, Micheal J. Beeler, Barbara A. McDonald.
p. cm. — (Cognition and literacy)
Includes bibliographical references and index.
Contents: v. 1. Programs, policy and research issues — v. 2. Theory and research in cognitive science.
ISBN 0-89391-737-0
1. Early childhood education—United States. 2. Cognition in children—United States. 3. Literacy programs—United States. I. Sticht, Thomas G. II. Beeler, Micheal J. III. McDonald, Barbara A. IV. Series.
LB1139.25.I58 1991
370.15′2—dc20 91-17779
CIP

Ablex Publishing Corporation
355 Chestnut St.
Norwood, NJ 07648

Table of Contents

Preface

Today there is widespread concern that the major tool for transmitting knowledge to new generations—the public schools—is failing. Numerous reports have noted that the dropout rate is high and among those who drop out, there is a low level of cognitive achievement. Even among those who graduate from high school, it is estimated that 1 in 6 may be "functionally illiterate." Low literacy levels are related to unemployment, welfare, and poverty. These problems are considered to contribute to low productivity in the workplace, as well, which is a large concern as the United States appears to be losing it's competitive edge in the world marketplace.

Education problems loom even larger as the future is considered. Demographic projections show that the population of youth is declining, except for minority groups, the very group that constitutes the largest percentage of dropouts, unemployed, and welfare and poverty groups. As this trend materializes, the future workforce will be comprised of people with lower cognitive skills. Predictions of work in the future show that while many jobs will be available in the lower-level, low-paying range, many jobs will be created in accordance with advances in new technologies. Fewer young adults will be available with cognitive ability needed to keep the U.S. competitive in the marketplace, and there will be an ever-widening gap between the haves and the have-nots.

In the face of these educational and demographic trends, there have been widespread calls for education reforms, lifelong learning, retraining of workers, and, in general, a concern for the reinvention of education in America to improve the cognitive abilities of children, youth and adults.

Given the widespread concern for improving the nation's cognitive ability, it was decided that perhaps the relatively new field of cognitive science might offer new perspectives that would aid in the development of new, more effective interventions. Consequently, a *Conference on the Intergenerational Transfer of Cognitive Skills* was convened in April 1988 in San Diego, CA. The participants in the conference reviewed the results of previous intervention programs for children, youth, and adults to determine if simply expanding on present efforts might make a significant improvement in the nation's cognitive ability. Additionally, they discussed a wide range of issues regarding the nature of cognition, how cognitive ability develops, and what educational implications can be derived from this scientific study of the mind.

The present book is one of two volumes of papers that were commissioned for the San Diego conference. Volume I reviews intervention studies of the last quarter century, and focuses on the emerging *intergenerational literacy programs* to explore the policy, practice, and research implications of such programs for meeting the nation's educational needs.

The present book, Volume II of the conference papers, presents a review of contemporary cognitive science and its implications for the intergenerational transfer of literacy and other cognitive skills.

The conference on the Intergenerational Transfer of Cognitive Skills and preparation of the books resulting from the conference was supported by a grant from the John D. and Catherine T. MacArthur Foundation to the Applied Behavioral & Cognitive Sciences, Inc. Much appreciation is given to Dr. Peter Gerber of the MacArthur Foundation for his encouragement and support of this project.

Thanks are also due to Mike Beeler and Debra Vella for their help in the conduct of the conference on the Intergenerational Transfer of Cognitive Skills and the preparation of materials for this book. Thanks are also given to Bill Armstrong, Richard Flyer, and Jane Wycoff who took copious notes at the conference and provided those notes to the editors for their use in preparing this book.

While acknowledging the help and assistance of these many colleagues, it should be noted that the opinions and conclusions given in this volume are those of the editors and authors. They do not necessarily represent the opinions or positions of the John D. and Catherine T. MacArthur Foundation or the Applied Behavioral & Cognitive Sciences, Inc.

Chapter 1

The Intergenerational Transfer of Cognitive Skills

Thomas G. Sticht
Barbara A. McDonald

It is well established that many children arrive at kindergarten or first grade with knowledge, language, and cognitive skills that are different from those needed to acquire higher levels of literacy, mathematics, and critical thinking abilities within the cultural context of mainstream public education. These children frequently fall behind in school and later drop out. They become the marginally literate and marginally employable youth and adults who comprise some one-fifth to one-third of the adult population in the United States. As many as 40 percent of minority populations may fall into this group (Sticht, 1987).

Many of these young adults become parents of children and are unable to transmit educationally relevant preschool oral and written language skills, which are the foundation for later reading and writing skills, or to model reasoning and thinking skills, frequently using mathematical concepts. Without these basics in language, numeracy, and critical thinking, the children from these homes show up for school prepared to recapitulate the failure of their parents, and the cycle repeats itself (Sticht, 1983; Hess & Holloway, 1979).

Attempts to break this cycle have focused on early childhood and elementary school-based interventions. Chapter I of the Education Consolidation and Improvement (formerly Title I of the Elementary and Secondary Education Act initiated in 1965) for 1986 included appropriations for some $3.5 billion for childhood interventions. In comparison, efforts to further develop the literacy, learning, and thinking skills of adults have been and are trivial. The total effort in adult literacy in federal programs has recently been estimated at some $347.6 million. But these dollars are spread across 14 federal agencies, and most are for direct services such as adult basic education or general educational development to obtain a high school equivalency certificate, or perhaps for learning English as a second language (Kahn, 1986).

EARLY CHILDHOOD INTERVENTIONS INVOLVING PARENTS

For decades it has been recognized that more literate, better educated, more highly intelligent parents somehow transfer cognitive skills to their children. In the early decades of this century, it was noted that more intelligent parents produced more intelligent children who achieved better in school and went on to become the better educated and achieving adults. It was thought that intelligence, defined operationally as the IQ, was largely an inherited and unchangeable trait that parents transmitted genetically to their children.

In the late 1950s, however, evidence was amassed to suggest that most of an adult's intelligence was developed by the time he or she was only 5 or 6 years old (e.g., Bloom, 1964), and that experience during this early period of childhood could be expected to change the child's IQ (e.g., Hunt, 1961). These ideas lead to the emphasis, during the War on Poverty of the mid-1960s, on early childhood interventions to break the cycles of low intelligence that lead to low school achievement, which, in turn, was thought to lead to low employability and poverty-level employment.

In several of the early childhood intervention programs attention was focused on trying to educate parents (and particularly mothers) in ways to stimulate the intellectual growth of their children. In a review of 17 early compensatory education programs, Bryant and Ramey (1987) conclude, not only that such programs can provide intellectual benefits, but that "the most improvement in intellectual development is seen when children attend daycare and families receive parent training or other service" (p. 71).

Today, the movement in early childhood intervention is away from working only with children in educational facilities, toward working with families in home, school, and community settings (Zigler &

Weiss, 1985; McCubbin, Olson, & Zimmerman, 1985; Klerman, 1985). Yet many important questions remain unanswered about how IQ builds, how parents facilitate this growth, and how to best teach both adults and children.

A NEED FOR KNOWLEDGE SYNTHESIS, DISSEMINATION, AND GENERATION

Though the convergence of adult and childhood intervention programs on the family as the basis for action is a promising trend, there are some serious problems that must be overcome if such programs are to have a consequential and measurable effect on the cognitive skills of both adults and their children.

A major problem is that the knowledge base on which to develop programs that aim to improve both adult's cognitive skills and such skills in the adult's children is very meager, if not nonexistent altogether. Understanding of adult cognitive skill development, particularly among the least able, is very limited. Unlike childhood cognitive development and education during the school years, where well over a billion dollars have been spent in the last decade and a half on basic and applied research, next to nothing has been spent to further our basic knowledge of the cognitive development of adults, particularly undereducated, out-of-school youth and adults.

To be sure, there have been billions of dollars put into job training and creation efforts, such as the Comprehensive Education and Training Act (CETA) and current Job Training Partnership Act (JTPA) programs. Unfortunately, these efforts have not supported basic research on the cognitive development of people in their programs. Repeatedly, program after program of youth remediation and job preparation has been developed, implemented, and soon discarded for lack of achievement of substantive improvements in learning. Furthermore, literacy programs for adults are lucky to retain adults for 50 or so hours of education. The residential Job Corps program attracts students for only up to 90 to 180 days; by then, most have left. Typically, only 1 or 2 "years" of growth in reading or mathematics is achieved, with nothing said of writing or critical thinking. And even the growth reported is suspect due to the use of inadequate assessment tools.

Outside of the military services, little systematic research has addressed the learning abilities of undereducated youth and adults (Sticht, Armstrong, Hickey, & Caylor, 1987). And, though some small-scale projects have suggested that the education of parents in such "parenting" activities as reading to their preschool children may bring about improvements in their children's achievement in reading in school (Maughan & Rutter, 1985), research has not been found on the extent to which adults can be educated to improve their own cognitive ability and how the new competence of the parents can be transferred to their children (Sticht, 1983).

The need for carefully planned and executed research in this area is indicated by the fact that recent testimony before the U.S. Senate Committee on Labor and Human Resources called for policy and funding in the hundreds of millions of dollars on the basis of a belief that "each grade more skills of parents in one generation yields an equal increment in the skills of the next generation" (Taggart, 1987, p. 28). The recommendation calls for the large-scale extension of a particular education program to bring about at least one year gain in adult basic skills, with the expectation that this new competence of adults would transfer intergenerationally to the adult's children . . . with no mechanism for this transfer planned nor elucidated.

A further problem with research on cognitive development, especially among undereducated youth and adults, is that it has typically been short-term research, accomplished by researchers who have come into a community, conducted the research and then left. Indigenous capacity for continuing the research in the community-based projects that are at work by the thousands throughout the nation is not developed and sustained. There are only limited efforts by "outsiders" to "put research into practice" without the long-term, sustained enquiry needed to usefully understand the functional contexts of adults in the local area and to improve the capacity of local organizations to design learning environments that promote cognitive growth in both adults and their children. This has the effect of failing to inculcate even the most rudimentary respect for measurement and evaluation in such programs. This shows itself in many ways, including a disdain for the collection of program outcome data, misuse of test instruments, and failure to learn from mistakes (Sticht, 1988).

Though, as indicated above, there is action in the policy, program, and research arenas to establish new models for the interruption of the intergenerational transmission of poverty and low cognitive skills, these activities are limited, and many are just forming. There has not been a concerted effort to bring the knowledge obtained in the last 20 years of childhood interventions to bear on the newly emerging adult–child interventions. There is still controversy over the success of the childhood interventions (Glazer, 1986; Jencks, 1986). Among other things, this includes concerns over the shift from reliance on the IQ as the central measure of cognitive growth in childhood programs to the use of school-oriented achievement tests (reading, mathematics) or noncognitive factors such as being promoted rather than held back, fewer delinquency problems, and so forth.

In adult cognitive skills remediation, little attention has been paid to the findings of the late 1950s and early 1960s regarding the rapid growth of IQ up to age 5 or 6 and then the relative stability of the IQ over the lifespan (e.g., Honzik, 1986). This finding is what suggested that most of the problem of adult illiteracy or functional illiteracy should be addressed through prevention in early childhood interventions rather than through remediation in adult literacy and basic education programs. What is the status of this IQ controversy today, and how does it relate to the move toward combined adult–child intervention programs? How do the new conceptions of intelligence (e.g., Bob Sternberg's Triarchic Theory, Howard Gardner's Multiple Intelligences, Olson's conception of intelligence as "the forms of competence involved in dealing with cultural artifacts or technologies" 1986, p. 338) relate to the current movement of cognitive intervention programs toward the family as the unit of action?

Not only are conceptions of intelligence different today from 20 years ago; additionally, conceptions of the so-called academic "basic skills" are different. Reading, writing, arithmetic, and higher order critical thinking skills are differently understood today than a quarter-century ago. A very significant difference in the understanding of these processes is the role of knowledge. Today, as contrasted with 20 years ago, there is much more of a recognition of the role of what one knows, in a very specific sense, in one's ability to apply the "basic skills" to problem solving and task performance in and out of school. What are the implications of the understandings from the recently emerged, interdisciplinary field called *cognitive science* for policy, practice, and research in adult–child interventions?

REVIEW OF RESEARCH

Because of the significance of the current move toward a new approach to the intergenerational transfer of cognitive skills that is taking place in the field of adult literacy development, including the proposed expenditures of hundreds of millions of dollars in federal and state education funds, it is both timely and important to conduct an in-depth review of the policy, practice, and scientific issues involved in such interventions.

As a part of a larger project, a series of scholarly papers by nationally prominent researchers and practitioners, some with experience in policymaking at the federal level, were commissioned.

The papers were presented and discussed in April 1988 at a 3-day *Conference on the Intergenerational Transfer of Cognitive Skills*. The conference included a review of past attempts to influence the course of cognitive development. Most of these programs were initiated in the mid-1960s as a part of the War on Poverty and were targeted at improving one generation at a time.

The papers for this part of the conference are presented in Sticht, Beeler, and McDonald (Volume I). Papers that describe ongoing programs to improve the cognitive skills of two generations, parents and their children, are presented in Part II of the same volume. Part III presents the summaries from the working groups regarding policy, practice, and research issues involved in the intergenerational transfer of cognitive skills. The book concludes with a brief, general summary and comment on the conference by Dan Wagner.

REVIEW OF THIS VOLUME

This volume presents a review of contemporary knowledge from the cognitive sciences regarding theories of cognitive development, intelligence, higher-order thinking, motivation, literacy, and cultural influence on cognitive development.

Cognition, Intelligence, and Thinking

In Chapter 2, Gavelek reviews selected theories of cognitive development and sets the stage for a new view of cognitive development by presenting it as a social phenomenon. This social view of cognitive development was advanced by Vygotsky in the 1950s but was largely unnoticed by American psychologists until recently. It stands in sharp contrast to the commonly held view of intelligence as an individual capacity. In Vygotsky's theory, cognitive development takes place within social interactions with caretakers and is governed principally by the language of the culture.

Gavelek's chapter contrasts the view of cognitive development held by Vygotsky with two other influential models of human cognition: Piaget's developmental model and the information processing model of cognitive science. Both of these models have advanced our understanding of cognition by emphasizing the constructive nature of cognitive processing. Piaget's theory has been the most important developmental perspective guiding early childhood education, yet even in this model the social role is facilitative but not formative. The information processing model provides a description of the system that processes knowledge but has no integrated description of the way this system develops. For the most part, there is not a social role attached to information processing. From a review of these theories, it appears that the inclusion of Vygotsky's model into the thinking of American psychology offers a way to incorporate the intergenerational transfer of cognitive abilities.

Wagner, in Chapter 11, presents a very complete discussion of intelligence. He looks at intelligence from the past, the present and what he suggests will be the future conceptions of human cognitive ability. In doing so, he, like Gavelek, describes the positions held by the psychometricians who were concerned with the structure of intelligence, the Piagetians who were concerned with structure and function, and information processing theorists who were working on the cognitive processes and the representations being processed while performing intellectual tasks.

Wagner presents some of the limitations of these traditional views of intelligence—the IQ-type tasks that represent a narrow subset of skills and the poor integration with learning and education. He goes on to discuss the present conceptions of cognitive ability which have gone beyond traditional IQ and now consider "practical intelligence." In his discussion of the future he suggests that there might be a return to interest in learning—in the capacity to learn. As he says, theoretically, learning and intelligence are closely related but there has been little work to specify the role of learning in intelligence and intelligence in learning.

In Chapter 12, Means critiques Wagner's paper and says that currently held views of practical intelligence show how narrow the traditional conception of intelligence has become. The need to define practical intelligence in a way that can be tested is stressed. She describes an aspect of practical intelligence that is found when studying expertise in a given field and provides a very interesting description of expertise among electronics technicians. She talks about the knowledge they have in order to solve various problems in troubleshooting, the ways in which problems are defined and the ways that they are solved. A large part of the discussion involves the concept of transfer as an indication of intelligence. Means says that being able to generalize from one type of problem to another is an important aspect of practical intelligence, in which there are systematically observable individual differences.

Chipman discusses another aspect of cognition in Chapter 9, that of high-order cognitive skills—what they are and how they might be transmitted. As Chipman states, we all have some idea of what is meant by higher-order skills, and the school system has been criticized for not producing students that can reveal these higher-order skills in their thinking, writing, and problem solving. But Chipman goes beyond the common sense notion of higher order skills and attempts to carefully define them in relation to various disciplines such as mathematics, logic, writing, and history.

She discusses higher order skills in the context of the functional architecture of human cognition that is prevalent in cognitive science research today. This is a complete review of the notions of conditions under which learned responses are likely to be emitted. Other topics include the limitations of working memory and the effects of practice on automatization of skills and effects on memory, age-related changes in memory, and task design to assess higher order skills separate from memory constraints.

Chipman relates these topics to the controversial questions of whether cognitive skills are domain specific or can be applied across a variety of tasks. In this discussion, she presents examples in problem solving, knowledge acquisition, and self-management skills. Finally, she presents an interesting discussion of how higher-order skills are transferred from one generation to the next, whether by parents or by formal schooling.

In a discussion of Chipman's paper, Nickerson (Chapter 10) agrees that the definition of higher order skills is an important one and a difficult one to make. He suggests that the distinction of higher- and lower-order skills can somewhat oversimplify the notion of human cognition. Using one of Chipman's own concerns, he discusses the implication for education offered by the idea of automatizing lower-order skills before higher-order skills are developed. In this example, he suggests that there are many individuals who show proficiency in a skill in spite of a lack of lower-order skills—writers who can't spell and mathematicians who can't compute without a calculator.

In discussing the difficulty of representing skills hierarchically, Nickerson offers the suggestion that, instead of trying to divide skills into higher- or lower-order categories, we instead make the distinction based on the manner in which various skills are executed. He discusses whether or not there are general cognitive skills or if there are only domain-specific skills. Various problem-solving heuristics and ways that they might be applicable in a number of problem areas are discussed. But again, Nickerson argues against classifying skills into either domain specific or general, and talks about thinking about skills in terms of their range of applicability to various problems.

Understanding the cognitive processes involved in thinking is only part of the problem. Motivational considerations must be raised as well. Bempechat, in Chapter 3, maintains that cognitive skills are socialized, which requires an understanding of the motivational skills as well. As she points out, cognitive skills can be present but not utilized effectively unless the right conditions are also present. According to Bempechat, the right conditions are ones that foster persistence in the learning environment, a preference for challenging tasks, and high expectations for eventual success. These conditions are influenced by the individual's beliefs about the nature of intelligence, whether it is believed to be a fixed or malleable condition. If a person believes that intelligence is a fixed capacity, attempts at learning that result in failure do not lead to further attempts. If, on the other hand, intelligence, is viewed as malleable or changeable, initially unsuccessful attempts to learn a new task are considered to be "okay"—practice will improve the performance. These beliefs lead to persistence or not and influence the choice of task and expectations for success.

Bempechat further points out that these beliefs can be manipulated. She shows how they are currently being manipulated to the disadvantagement of many minorities, including females. She provides a discussion of the ways that parental beliefs manipulate childrens' views of themselves and how this influences school abilities such as math learning. The attributional style of Asian children is offered as an adaptive belief system with accordingly positive results.

Mathematics and Literacy

Many of these motivational issues are explored with respect to parents' influence on childrens' mathematical achievement in Chapter 8 by Ginsburg, Bempechat, and Chung. In this paper, cultural differences in mathematical achievement are presented and confirm our fears about U.S. schools in relation to Asian schools in terms of mathematics. Alarming statistics are presented, such as the fact that, in one study of mathematics achievement among fifth graders, the highest U.S. score was lower than the lowest Japanese score. Ginsberg, Bempechat, and Chung build the argument that motivation and attitudes have a lot to do with the difference in educational achievement, not just between U.S. and Asian children but between social classes in this country, as well.

In attempting to account for differences in mathematics achievement, the authors show that it is not genetic differences alone, it is not simply differences in teaching styles, and it is not that mathematics is considered a more difficult subject by U.S. students (interestingly, in one study U.S. students thought mathematics was not so difficult but disliked it, while the Asian counterparts thought mathematics was very difficult and they also disliked it). The cultural framework surrounding the development of Asian children's mathematics achievement is one that stresses the need to adapt to difficulty, to persist in the face of difficulty and to view ability as always present. Effort is critical to performance and is something that can always be improved. This is a fascinating comparison of cultural approaches to mathematics and compliments Bempechat's earlier paper on motivation.

In a critique of Ginsberg, Bempechat, and Chung's paper, Marshall (Chapter 8) takes the concepts back into the U.S. culture and suggests that, given the need to help parents become better facilitators of childrens' mathematics learning, we should identify which skills should be transferred. As Marshall says, we do not expect parents to be surrogate mathematics teachers—most parents, especially those of low-achieving students, do not themselves have strong mathematics skills. She agrees with Ginsberg et al. that the expectations and beliefs that surround mathematics are important to childrens' performance and need to be positive. She talks about the importance of supervising study, being the role models in persistence, conveying that mathematics is valuable, and working with students in solving everyday problems that use mathematics skills.

Marshall points out, however, that parents need to be given some training in order to be appropriate role models. Many parents do not see the value of mathematics in their own lives and their own mathematics problem solving does not seem to relate to what their children are doing. This is where Marshall thinks attitudes can be changed. It is not that parents need to show that they, too, could pass calculus, but rather that they should foster a problem-solving environment in which the mathematics skills that are used are made explicit. She gives some excellent suggestions of ways to work together as a family on long-range projects that require planning, measuring, mathematics computation, and other valuable mathematics skills.

Issues that arise when efforts are made to implement programs to change cognitive ability in large groups of children are discussed in Chapters 4, 5, and 6. These issues include the theoretical assumptions guiding program development, the problems of intervention, not to mention the politics of intervention, and data collection and interpretation.

In Chapter 4, Mason and Kerr discuss the components of literacy acquired during preschool years and focus on phoneme and grapheme awareness, conceptions of print, and written language features. They then discuss how parents can transfer literacy skills to their children through understanding of these components. They discuss successful transfer which makes use of more explanation, teaching the ABCs, playing school, and storybook reading—things that parents with good reading skills probably do spontaneously.

In discussing the negative cycle that is set into place by sociocultural factors, they carefully distinguish between social class and culture and ask, "What kinds of intervention can break the cycle of low literacy in adulthood perpetuated into the next generation?" Intervention programs are presented that show promise in promoting intergenerational transfer and that also demonstrate some very interesting problems to be addressed in the development of such programs. In concluding, Mason and Kerr address an additional issue which bears further research. They suggest that, by ignoring the values of orality, Western culture may have inadvertantly bred illiteracy in some minority groups which have a history of oral storytelling. They suggest that orality can be used to promote literacy instead of being replaced by literacy.

In a critique of Mason and Kerr, Slaughter (Chapter 5) presents the explicit and implicit theoretical assumptions guiding the work and suggests that, while it can be shown that home environment influences early cognitive development, many other assumptions are as yet untested in 1988. She challenges the idea that we know enough to properly design programs to remedy cognitive deficits—and, furthermore, wonders whether we know who would be targeted by these programs. As she clearly states, the ties between the scientific child development community and social and public policy are extremely tentative, which makes the implementation of current scientific work problematic.

In a discussion of the implementation programs cited by Mason and Kerr, Slaughter states that these results are not conclusive because there were no control groups and no way of disentangling the gains in cognitive development from the different social and cultural groups. The criticisms that she makes are very similar to those made in many areas of educational implementation and warrant our attention.

Cognition and Culture

In Chapter 6, Fox presents a complete and moving account of the attempts to integrate American Indians into the dominant society. In this chapter, Fox presents the tribes of North and South Dakota (the Sioux, Ojibwa, and the Three Affiliated Tribes) as the targets and victims of a bungled bureaucratic approach to "civilize" native Indians. Fox clearly plots the development of intervention programs throughout their history, placing them into historical context and showing how these programs moved back and forth from attempts to remove Indians from their cultural influence to attempts to view their culture as important—different but not deficient.

In providing this historical perspective, Fox also shows the educational assumptions that were in place at these points in time and how they influenced the programs and reception by Indians. The era of self-determination by Indians has extended from 1969 to the present, and yet the educational model is currently based on the no-difference viewpoint and not much "Indianness" has been included in the programs. This is an important paper for any educational implementer to read and absorb—it points out many potential problems that occur when intervention programs are designed and implemented, especially when the culture in which the program will be implemented is not thoroughly considered.

Finally, in Chapter 13, Reder presents a thorough discussion of the importance of cultural factors in intellectual development and helps tie together many of the issues discussed throughout the book. The concept of culture has become an important aspect of general descriptions of cognitive abilities (note the chapter by Gavelek, this volume, about Vygotsky's social and cultural factors) but as yet cultural factors have not been well integrated into cognitive science.

Reder discusses the socialization of specific types of intellectual behaviors—what are the settings and social interactions in which intellectual behaviors are socialized by others? By looking at the problem in this way, he says we can include many influences on cognitive development, from school to home to job to community. In looking at the influences of culture, Reder discusses the methods of observation of literacy practices of various groups and presents three socialization perspectives that offer interesting insights into the cultural transmission of literacy and also some ways in which its development might be stimulated. From all of the work that he presents, he draws some important implications about ways to encourage literacy development, such as incorporating the use of written materials into existing social practices. He talks about utilizing natural helping relationships, so that the helpers are trusted friends or relatives, by embedding literacy training activities into meaningful cultural contexts and by broadening the general contexts in which literacy development takes place. School is but one of those places.

REFERENCES

Bloom, B. S. (1964). *Stability and change in human characteristics.* New York: Wiley.

Bryant, D. M., & Ramey, C. T. (1987). An analysis of the effectiveness of early intervention programs for environmentally at-risk children. In M. J. Guralnick & F. C. Bennett (Eds.), *The effectiveness of early intervention for at-risk and handicapped children.* New York: Academic Press.

Glazer, N. (1986). Education and training programs and poverty. In S. H. Danziger & D. H. Weinberg (Eds.), *Fighting poverty: What works and what doesn't.* Cambridge, MA: Harvard University Press.

Hess, R., & Holloway, S. (1979). *The intergenerational transmission of literacy*

(Report prepared for the National Institute of Education). Washington, DC: U.S. Department of Health, Education and Welfare.

Honzik, M. P. (1986). The role of the family in the development of mental abilities: A 50-year study. In N. Datan, A. L. Greene, & H. W. Reese (Eds.), *Life-span developmental psychology: Intergenerational relations*. Hillsdale, NJ: Erlbaum.

Hunt, J. M. (1961). *Intelligence and experience*. New York: Ronald Press.

Jencks, C. (1986). Comment (on the chapter by Glazer). In S. H. Danziger & D. H. Weinberg (Eds.), *Fighting poverty: What works and what doesn't*. Cambridge, MA: Harvard University Press.

Kahn, M. E. (1986, May). *Literacy management information project report: Volume I* (Report for the Adult Literacy Initiative). Washington, DC: U.S. Department of Education.

Klerman, G. L. (1985). Community mental health: Developments in the United States. In R. N. Rapoport (Ed.), *Children, youth, and families*. Cambridge, UK: Cambridge University Press.

Levitan, S. A., & Gallo, F. (1988). *A second chance: Training for jobs*. Kalamazoo, MI: W. E. Upjohn Institute for Employment Research.

Maughan, B., & Rutter, M. (1985). Education: Improving practice through increasing understanding. In R. N. Rapoport (Ed.), *Children, youth and families*. Cambridge, UK: Cambridge University Press.

McCubbin, H. I., Olson, D. H., & Zimmerman, S. L. (1985). Family dynamics: Strengthening families through action-research. In R. N. Rapoport (Ed.), *Children, youth and families*. Cambridge: Cambridge University Press.

Sticht, T. (1979, May). *The cycles of marginal literacy and marginal living*. Paper presented at the Annual Meeting of the Association of American Publishers, Palm Beach, FL.

Sticht, T. (1983). *Literacy and human resources development at work: Investing in the education of adults to improve the educability of children* (HumRRO Professional Paper 2-83). Alexandria, VA: Human Resources Research Organization.

Sticht, T. (1987). *Functional context education: Workshop resource notebook*. San Diego, CA: Applied Behavioral & Cognitive Sciences, Inc.

Sticht, T. (1988). Adult literacy education. In E. Rothkopf (Ed.), *Review of research in education* (Vol. 15). Washington, DC: American Educational Research Association.

Sticht, T., Armstrong, W., Hickey, D., & Caylor, J. (1987). *Cast-off youth: Policy and training methods from the military experience*. New York: Praeger.

Sticht, T. G., Beeler, M. J., & McDonald, B. A. (Eds.). (1991). *The intergenerational transfer of cognitive skills: Volume I: Programs, policy and research issues*. Norwood, NJ: Ablex Publishing Corp.

Taggart, R. (1987, January). *Solving the basic skills crisis* (Testimony before the U.S. Senate Committee on Labor and Human Resources). Washington, DC: Remediation and Training Institute.

Zigler, E., & Weiss, H. (1985). Family support systems; An ecological approach to child development. In R. N. Rapoport (Ed.), *Children, youth and families*. Cambridge, UK: Cambridge University Press.

Chapter 2

Developmental Perspectives on the Intergenerational Transfer of Cognitive Skills

James R. Gavelek
Michigan State University

INTRODUCTION

Solutions are oftentimes easy. It is the framing of the problem that is the problem. The problem before us in this conference is ill-defined. Not through any fault of the conference organizers, but because concepts such as undereducation, poorly developed cognitive skills, illiteracy, and, I would add, learning disabilities, are themselves concepts that are ill-defined and fuzzy. Furthermore, they are concepts heavily infused with social meaning—meaning that can vary cross-culturally and change historically.

Even the notion of intergenerational transfer of cognitive skills suggests a certain ambiguity. There are two interactive mediums through which cognitive skills, or the disposition to acquire these skills, are transferred (or not transferred) intergenerationally—one biological and the other cultural. While analytically separable, it is clear that in reality nature and nurture interact in ways that at times appear intractable. One thing that cannot be disputed, however, are the costs, both psychological and financial, that the problems mentioned above exact upon society, its individuals, and their loved ones. They are staggering!

An important issue in conceptualizing intergenerational learning problems is their cyclical nature. Individuals who have not had the appropriate guidance in their learning as children are not likely to provide appropriate guidance for their own children. Indeed, one of the most serious problems confronting poor adult learners may be that they know neither what nor how to foster effective learning in their own children. How many such problems be prevented in children or, failing prevention, be remediated once children have achieved adulthood? For me, this is the major orienting question of the conference.

The notion of intergenerational transfer of cognitive skills (ITCS) speaks to the fundamentally social nature of mind. Implied is the idea that both what is transferred and how this transfer takes place have their origins in the relationships between people. The present chapter addresses the role that theories of cognitive development might play in informing our understanding of intergenerational transfer. More specifically, I shall attempt to evaluate the extent to which the content and mechanisms of ITCS are coextensive with those of cognitive development. Toward this end, I examine three major theories of cognitive development: Piaget's structural-developmental, information processing theories in cognitive science, and Vygotsky's sociohistorical theories. Questions pertaining to intergenerational transfer are by their very nature life-span in scope. A second major purpose of this chapter is to evaluate the status of the above-mentioned theories as life-span conceptions of development. In effect, what is being evaluated is the extent to which these theories entail a social-cognitive conception of mind in both children and adults.

There have been literally hundreds of volumes written about each of the three theoretical perspectives that will be discussed in this chapter and I cannot present all that I would like. This presentation in this chapter will be structured around the dual concerns of the what and how of both ITCS and development. Thus, one may inquire as to what is transferred when intergenerational transfer of cognitive skills takes place and ask the parallel question, what is developed through cognitive development? Similarly, the question of how intergenerational transfer takes place may be posed in parallel with questions regarding the mechanisms of cognitive development.

I begin with an analysis of the concept of intergenerational transfer as it relates to cognition. It is suggested that "unpacking" this concept involves the consideration of a number of important dimensions. What is transferred between generations, between whom, in what contexts, and how? Do the answers to these questions have to be qualified depending upon the historical period of time (or culture) one is examining? These are issues of the utmost importance if we are to make sense of the problems at hand.

Next, I turn to theories of cognitive development and suggest two general criteria by which they may be evaluated. As mentioned above, what is developed through cognitive development? How does this development take place? After elaborating these questions I then turn to a consideration of each of the above-mentioned theories. After briefly overviewing each theory I apply the "what" and "how" criteria. I conclude this section with a summary of the extent to which these theories address: (a) the conceptual overlap between ITCS and cognitive development; and (b) the extent to which these theories offer life-span accounts of the role intergenerational transfer in cognitive development.

Third, I argue that we give serious consideration to an additional theoretical orientation—understanding the development of our implicit or common-sense conceptions of cognitive development. In so doing, I suggest the necessity of recognizing the ideology that pervades our conceptions of cognitive development. The need to do so is as great for those of us who are social scientists, policy makers, and practitioners as it is for the public(s) we seek to serve. It is suggested that in the absence of such knowledge we cannot thoroughly understand why it is that children or adults "own," or fail to own, the cognitive goals and means for achieving those goals that society expects that they acquire.

I conclude this chapter by considering some of the major implications of developmental theory for the conceptualization of research issues, policy decisions, and practice. These implications are considered with respect to problems in the acquisition of cognitive skills in children and adults.

THE INTERGENERATIONAL TRANSFER OF COGNITIVE SKILLS — FRAMING THE ISSUES

What is the role of the intergenerational transfer of cognitive skills in the development of mind? Consider the following thought experiment:

> A highly contagious but short-lived disease sweeps over the earth and in a period of weeks all of its adult population perishes. The only survivors are very young infants who because of their temporary natural systems of immunization are unaffected by the disease. Fortunately, because of advances in robotics the biological needs of these infants (e.g., hunger, thirst, warmth, and so forth) have been provided, and the survival of most of them is insured. Moreover, the infants are reared in groups thus guaranteeing the opportunity for social interaction. Consider both the

> short- and long-term consequences of this adult-free world for the intellectual development of those individuals who do survive as well as for those of the generations to follow. (Gavelek, 1986, p. 5)

This thought experiment focuses on the social nature of learning. As Cole (1988) reminds us, a special property of humankind as a species is that we are able to regulate our actions through culture. Cultural artifacts in the absence of those knowledgeable individuals to mediate their meaning are meaningless. Individuals who have grown without the benefit of cultural artifacts are cultureless. It is through the evolution of educability that we have become able to deliberately teach and learn, both formally and informally, and in so doing perpetuate and at the same time transform our culture.

Like proverbial fish we move about in a sea of cultural artifacts, the pervasive influence of which we do not see. The thought experiment above suggests, in the starkest of terms, the role that knowledgeable representatives of the culture play in mediating our actions without our world. Remove these representatives, and what we are left with, at least for some time to come, are individuals whose understanding of their world is episodic and incoherent.

What is meant by the intergenerational transfer of cognitive skills? The *Random House College Dictionary* defines a generation as "the term of years, accepted as the average period between the birth of parents and the birth of their offspring" (p. 550). I take this to mean that a generation in our society is somewhere in the neighborhood of 25 years; but with considerable variability owing to the wide range of parenting ages.

On its surface, the meaning of ITCS would seem obvious. It refers to the transfer of knowledge concerning the use of cognitive skills between members of different generations. However, there are a number of questions suggested by the concept that need to be considered before evaluating the role that developmental theories may play in informing our understanding of the ITCS. One way of understanding the concept is to ask what is transferred, in what contexts, between whom, and how?

What are the contexts in which intergenerational transfer takes place (e.g., at home, in the schools, on the job)? Between whom are cognitive skills transferred (e.g., parent and child, teacher and student, craftsman and apprentice)? The assumption implicit in the concept of ITCS seems to be that members of one generation (e.g., adults) may know something of importance (i.e., ways of knowing) from which members of another generation (e.g., children) might benefit. Implied is a certain assymmetry between one generation, typically older, members of which are "in the know" relative to those of another, younger generation who are less knowledgeable. It is important to recognize, however, that the concept of a generation represents a potential confounding of knowledge with cohort groups. Thus, it is conceivable that younger children might acquire certain knowledge (e.g., the use of computers), that their elders have not. The processes of ITCS may be reversed. Thus, is it the generational differences, the knowledge associated with a given generation, or both, that is important?

Do we acquire different knowledge concerning cognitive skills from other generations than we might from our own? In what important ways does *inter*generational transfer of cognitive skills vary from the transfer taking place among peers or members of the same generation (i.e., *intra*generational transfer).

What—if anything—is distinctive about how learning occurs between individual members of different generations? What are the mechanisms by which knowledge of cognitive skills are transferred between members of different generations (e.g., are they taught, acquired through observation or by incidental learning)?

Any interpretation of ITCS has to be set in both cultural and historical contexts. Thus, the cognitive skills transferred, the contexts in which this transfer takes place including both the setting (e.g., home) and participants (e.g., parent and child), and the mechanisms by which this transfer is brought about (e.g., school taught learning) may all vary across cultures and over extended (i.e., historical) periods of time.

There is good reason to believe that the cognitive skills transferred between generations today are different than they were in other periods of our history. Toulmin (1979) suggests that reading as a silent activity is a relatively recent invention. Up until the 15th century, to read was to read aloud. Others have documented the changes brought about in memory skills with the invention of the printing press. A comprehensive accounting of the natural history of our cognitive processes has yet to be written.

Similarly, there are historical differences in both the contexts in which we acquire cognitive skills as well as the individuals from whom they are acquired. Thus, we have become a more schooled society and it has been from members other than one's family that we acquire what it is that we know.

Finally, the processes or means by which we acquire cognitive skills would seem to have changed historically as well. The transition from orality to literacy has meant changes not only in what we learn but how we learn. Moreover, becoming a more schooled society has meant that we must learn and think in ways that are more decontextualized or spatially and temporally remote.

THE DEVELOPMENT OF COGNITION

Developmental Criteria

A useful way of evaluating how developmental theory might inform our understanding of the ITCS is by assessing the extent to which the content and mechanisms of cognitive development are coextensive with those of ITCS. In other words, does development and how this development takes place include, at least in part, the content and mechanisms of ITCS? Accordingly, I have chosen two major overarching questions by which to evaluate the developmental theories to be discussed. The first question relates to the what and the second to the how of development. Each of these issues entails its own set of associated questions. It should be emphasized that such a distinction is for analytic purposes alone. We shall see that in point of fact, it is often difficult to decouple what is being acquired from how. Other than as an abstraction, content does not exist apart from process, nor vice versa.

What develops through the development of cognition? De Gelder (1986) has argued that cognitive development is a difficult notion to explain primarily because of the ill-defined nature of the concept of cognition. She suggests that at the core of the idea of development there is an assumption of a specifiable pre-existing state, an end state, and a set of necessary sequences that join the former to the latter. De Gelder maintains that we need to understand what is it that goes into these pre-existing and end states before we turn our attention to consideration of the nature of the developmental mechanisms by which one is transformed between them. We need to understand what it is that develops when development occurs.

The concept of cognition traditionally has been used in at least two important ways. First, it has been used as an adjective describing a particular class of psychological (i.e., cognitive) processes. Thus, cognitive processes refer to the functions by which an individual comes to know the world. Here, cognitive development refers to the development of processes consisting of activities such as problem solving, remembering, or language. The second sense in which cognition has been used is as a synonym for the product or outcome of the processes of knowing (i.e., knowledge). By this criterion, cognitive development is essentially the study of the acquisition of knowledge. Both of the above senses of the term cognition, however, require further specificity. Thus, one may pose both the questions of "what characterizes a process as cognitive?" and "what is meant by knowledge?" These are not questions to be pondered by philosophers alone. Their answers have an important bearing on how we think about what it is that is transferred between generations. The analytic distinction between the meaning of cognition as both process and product is nicely framed in a question raised by Shaw and Bransford (1977), "What is the nature of the epistemic process by which people know their world and the nature of the world that it might be known?" (p. 11).

Issues concerned with the meaning of knowing and knowledge are enormously complex and entail assumptions about the individual as a knower, the world as known, and the nature of the relationship between the two. The individual influences the world through his or her actions (i.e., purposive behavior), but at the same time is influenced by the world through his or her perception of it. The relationship is dialectical.

The following issues become important in assessing the role of intergenerational transfer in the development of cognitive skills. To what extent does the world to-be-known consist of cultural artifacts, the meanings of which are societally determined, culturally variable, and historically changing? And, are the purposes and operations constitutive of an individual's actions on his or her world, and his or her perceptions of that world, socially mediated? Is the role of a mediating agent merely that of facilitating the development of processes intrinsic to the individual? Or is the function of the mediator to transmit ways of knowing and knowledge that would not otherwise occur in the absences of such mediation? In short, to what extent is our knowledge of the world, and the processes by which that world is known, social constructions? (Gavelek, in press; Harre, 1986; Mehan, 1981; Gergen, 1980, 1985).

How does development occur? Predictably, most developmentalists argue that an understanding of the mechanisms of development is the more important of the two issues, despite the fact that it is an area that has been given relatively little attention.

Clearly implicated in the concept of development is the notion of change; and yet change is something that the concept of learning is about as well. To what extent do theories of development distinguish between the concepts of learning and development? The following criteria have been generally used to distinguish between learning and development:

1. Learning refers to more episodic changes taking place over shorter periods of time. Development suggests a more protracted process involving structural reorganization.
2. The processes of learning are assumed to be reversible; those of development, much less so.
3. Change attributable to learning is generally assumed to be the result of factors acting on the individual from without, while those attributable to development emphasize determinants arising within the individual.
4. Changes attributable to learning are usually specifiable in quantitative terms while developmental change has been thought of as more qualitative in nature.
5. Changes attributable to learning are generally assumed to be more specific while those attributable to development more general.
6. There is an assumption of progress implicit, if not explicit, in the concept of development that one does not find with respect to learning. One develops toward a better, more organized, functional, competent individual. One may learn habits—good or bad, functional or dysfunctional. While they are distinguishable, the concepts of learning and development are not unrelated. Too often the two have been treated as oppositional rather than as essentially complementary and related processes. Any comprehensive account of the development of cognitive skills must also address the role of learning in how these skills are acquired.

Second, how broad or narrow ought our conception of development be? Traditional conceptions of development have focused on ontogenesis or the development of the individual. To what extent need a theory address other developmental dimensions in addition to a person's individual or ontogenetic history? Baltes and Goulet (1970) have suggested five different developmental psychologies that may assist our understanding of the problems associated with conceptualizing the role of development in the ITCS. These psychologies correspond to the study of age-related changes between different (a) species, (b) cultures, (c) generations, (d) periods of a life-span, and (e) psychological systems (e.g., memory).

A final factor that must be considered in accounting for the mechanisms of development is the relative extent to which the locus for the developmental dynamic is assumed to reside within the individual (i.e., endogenous) and/or the environment (i.e., exogenous). At issue here is the concept of action. If action is to be interpreted as purposive behavior, then what are the origins of these purposes? Are individual's assumed to be active cognitive agents in that they construct their own purposes, or are they to be thought of as essentially passive—following purposes that are extrinsic? The particular motivational construct(s) that a theory invokes to explain the reasons for an individual's actions need to be made explicit. Traditionally, the concept of development has assumed a universal endpoint toward which we all develop. Kessen (1984) suggests that we examine the extent to which individuals march toward a common and unchanging development endpoint, or whether development may follow any of several different trajectories.

As suggested earlier, the degree to which developmental ends and means can be disaggregated is questionable (Kessen, 1984). Thus, while I have separated issues associated with what develops from those that relate to how this development comes about for analytical purposes, it should be stressed that the two are interrelated. The mechanisms assumed necessary for development are likely to be coupled with what it is that is assumed to develop (i.e., the unit of analysis). For example, cognitive skills believed to be more basic may also be more assumed to be more maturationally based than those that are thought of as higher and derived.

Piaget's Structural-Development Approach

Piaget rejected the idea that knowledge was simply a copy of reality (Piaget, 1954, 1984). Instead, he assumed that through actions carried out on objects and events, individuals are able to construct their own knowledge. To know an object is to act upon it, transform it, and thus come to understand how it is constructed. This occurs as the result of operations, that is, internalized actions which modify the objects of knowledge. What develops according to Piaget are schemata, organized structures of these operations by which the individual interacts with the world. Piaget (1964) believed that the central problem of development "is to understand the formation, elaboration, organization, and functioning of these structures" (p. 39).

Through development these psychological structures are transformed into increasingly rich and internalized operational structures that are organized in a synchronic, stage-like, fashion. In order to understand the psychological functioning of an individual one must understand the system of logic underlying their thinking at a given stage of development. Piaget identified four such stages: (a) sensory-motor, (b) pre-operational, (c) concrete operations, and (d) formal operations.

The origins of what will eventually become an individual's operational structure (i.e., schema) are to be found in infancy and early childhood during the sensory-motor stage. Here individuals first come to know their world through their externalized interactions with it. It is during the second or pre-operational stage where we observe the beginnings of symbolic representation including language. Piaget assumed that at this stage individuals must reconstruct all that was developed during the sensory-motor period, but now at a symbolic level. He believed that it is logic, specifically, logical-mathematical structures, that underly all thinking and that language plays a secondary or derivative role. As suggested by its name, the stage of concrete operations is characterized by individuals being able to engage in operational thinking, but in a way that is still independent of their powers of symbolic representation. At this level they can accomplish through their concrete actions what they cannot yet do more abstractly through the vehicle of language and other symbol systems. The highest stage of cognitive development—that of formal operations—is generally achieved during adolescence and is characterized by the ability to think abstractly and hypothetically.

Piaget equated knowledge with scientific knowledge. The endpoint or destination of cognitive development associated with formal operations is to achieve a level of thinking associated with modern natural science and mathematics. Russell (1978) maintains that Piaget's science-oriented epistemology leads to a neglect of the conceptual systems of everyday cognitive life in adults. Piaget recognized little development of importance beyond formal operations. Indeed, he expressed his position on the study of adult cognitive development rather clearly when he maintained that "for our purposes studies of adults are completely worthless . . . if one wishes to continue the study of stages beyond adolescence the only method would be to use history, for example the history of scientific thought."

Toulmin (1981) questions the extent to which cognitive development should be assumed to have a unique destination, asking,

> How far is it possible to determine and define in advance the final goal or destination toward which improvements in our procedures and abilities are directed, either during historical evolution of our collective procedures, or during the psychological development of our individual abilities?

Toulmin asks how it is that children come to "recognize the necessity of conforming to the intellectual structures of logic" and indeed, why there ought to be such a necessity. He suggests that the crucial question we should be asking is "in what respects, to what extent, and on what conditions, does the psychological development of human individuals — like the historical evolution of human cultures—proceed in fact toward some universal and uniquely definable destination?" What little evidence we have undercuts any supposition that intellectual evolution and conceptual development have a single, unique, and universal destination. The directions of conceptual ontogeny and phylogeny alike are to some significant degree indeterminate and open-ended. Toulmin concludes his critique by suggesting that for Piaget logical "necessity" (i.e., systematicity) was equated with pragmatic "necessity" (indispensability).

How does development take place? Although Piaget identified four mechanisms as responsible for development, not all occupied a central position in his theory. The mechanisms are: (a) maturation, (b) experience, (c) social transmission, and (d) equilibration.

Maturation. While strongly influenced by his background in biology, Piaget went out of his way to emphasize that maturation was only one of several factors responsible for an individual's development and that his was essentially an interactional theory of development.

Experience. Piaget differentiated between physical and logical-mathematical experience. He believed this distinction to be important both psychologically and pedagogically. We experience an object physically by acting upon and abstracting knowledge from it. In our logical-mathematical experience, knowledge is not drawn from the object but by our actions effected upon the objects. For example, in counting a series of objects we not only experience the objects in and of themselves (i.e., physical experience), but also discover properties associated with the act of counting itself (i.e., logical-mathematical experience). Piaget believed that it was action itself that was fundamental in accounting for intellectual functioning. The role of sign systems (e.g., language and mathematics) was derivative and secondary in importance.

Social transmission. By social transmission Piaget meant both linguistic and educational transmission. He was careful to point out, however, that a child can only receive valuable information to the extent that he is in a state where he can understand information. Social transmission along with experience (i.e., physical and logical-mathematical) address Piaget's conception of learning. However, Piaget (1964) conceived of learning as essentially subordinate to development. He maintained that

> A stimulus is a stimulus only to the extent that it is significant and it becomes significant only to the extent that there is a structure which

> permits its assimilation, a structure which can integrate this stimulus but which at the same time sets off the response. (p. 43)

Thus, whether one promotes children's understanding of their physical world through exposing them to manipulables or seeks to verbally instruct them in the acquisition of a particular concept, they will only assimilate this knowledge to the extent that there are operative structures (i.e., schemata) that permit such learning to take place.

Piaget questioned the extent to which development could be accelerated through instruction and suggested that three criteria had to be met before he would accept structural change as evidence of learning. The learning had to be: (a) lasting, (b) generalizable, and (c) result in an enhanced operational level of thinking relative to that which took place prior to the learning experience. Thus, Piaget assumed that generalizable knowledge did not, indeed could not, be taught but resulted from a widespread transformation in operational structures that characterize development.

Equilibration. It is equilibration of actions that was the cornerstone of Piaget's conception of developmental change. According to this notion, individuals are motivated to bring their operational structures (i.e., schemata) into equilibrium with what they experience in their environment. They may do this by either a process of accommodation of these structures to the environment, or the assimilation of the environment in a manner consistent with their present cognitive structures. Essentially the individual is motivated—epistemically—to construct his or her own knowledge but in so doing, avoids cognitive conflict or inconsistency. Piaget eschewed the use of motivational constructs other than that of equilibration to explain cognitive growth (Russell, 1978).

Piaget on the intergenerational transfer of cognitive skills. We have seen that, for Piaget, "what develops" is an individual's scientific and mathematical knowledge of the world. As such, Piaget tended to minimize or ignore the role of language in preference to an individual's actions and his or her resulting operational structures. He had little to say about the development of everyday concepts of the world. Moreover, Piaget's conception of the levels of mathematical and scientific knowledge that are developed appears frozen in time. While he recognized and discussed phylogeny or the history of the species, he doesn't seem to have incorporated this historical dimension into his accounting of the endpoints of individual development. What develops is assumed to be universal and highly general rather than historical and contextual.

While a constructivist, Piaget's conception of the acquisition of knowledge was one of individual construction. The major benefit to be gained by individuals—mainly peers—working together is motivational rather than cognitive in that the failure of two individuals to agree on a problem solution results in disequilibrium which is the impetus to further development toward universally held beliefs. Differences in what is known between each other, and thus might be shared with each other, were accorded little importance by Piaget. The role of the social is facilitative more than formative. Russell (1978) suggests that Piaget underplayed the role of social factors and that he gave only token attention to education and cultural transmission.

> Every process, including the social, is a manifestation of the universal tendency towards equilibrium within structures. Piaget regards society as a model of the organism, so it is not possible within Piagetian theory to talk of the cultural transmission of knowledge because this process is itself the working out of an equilibration whose principles transcend and underly social phenomena; "education" is taken to affect only the ease with which operativity is attained. (Russell, 1978)

Information Processing

It may seem strange to include information processing as a major theoretical approach to the study of cognitive development (Rabinowitz, Grant, & Dingley, 1987; Sternberg & Powell, 1984). We shall see that, in certain important respects, the reader's skepticism is warranted. However, it is important to point out that information processing has become the dominant paradigm in cognitive psychology (Simon, 1980) to the extent that while the relatively new multidisciplinary designation—cognitive science—is supposed to integrate areas such as psychology, linguistics, computer science, philosophy, and anthropology, for many psychologists it has become almost synonymous with information processing theory. Moreover, there are a number of prominant information processing psychologists who have identified themselves as developmentalists (e.g., Sternberg, Klahr, & Siegler). In what follows I shall try to characterize information processing approaches to cognition and evaluate the extent to which such approaches provide an adequate conception of the development of cognition.

Most information processing psychologists maintain that a more productive understanding of human mentality comes from discovering the "program" and "plans" underlying professional rather than underlying physiological representation (Teske & Pea, 1981). Toward this end, Winograd and Flores (1987) identify the following set of assumptions that underlie information processing approaches:

1. All cognitive systems are symbol systems. They achieve their intelligence by symbolizing external and internal situations and events, and by manipulating those symbols.
2. All cognitive systems share a basic underlying set of symbol manipulating processes.
3. A theory of cognition can be couched as a program in an appropriate symbolic formalism such that the program when run in the appropriate environment will produce the observed behavior.

In evaluating information processing as a conception of cognitive development, it is important to understand both what it is that is assumed to develop and how this development takes place.

What develops? In virtually all information processing models of cognition the existence of a set of elementary or primitive (i.e., innate) information processes are assumed from which all higher-order cognitive processes are built (Siegler, 1983). What develops are representational systems or states (i.e., an individual's knowledge). These representational systems may be modeled using various descriptive formalisms for characterizing the structures and processes that constitute an individual's knowledge. One may construct a composite picture of development by modeling and sequencing these knowledge structures at different development levels.

Langley (1983) characterizes the "cognitive architecture" of an information processing system as those components assumed to remain constant across individuals, task domains, and over time. There are a number of such systems that have been studied including rules, scripts, skill hierarchies, semantic nets, grammars, and production systems (Klahr, 1984). Research on production systems are particularly helpful in understanding an information processing conceptualization of development. Newell (1973) maintains that systems of these productions are all that is necessary for effective information processing.

Production systems consist of a series of rules each of which can be represented in terms of condition-action (i.e., if-then) statements. If a data pattern satisfies a condition of data already residing within working memory, then the action component of the production is set into operation which then alters the contents of working memory. In this way, productions are interactive with memory such that, once set into action, a production may then alter what is represented in memory. This, then, has implications for subsequent condition-action pairs, and so on.

How does development take place? Teske and Pea (1981) suggest a central issue in information processing conceptions of cognition is that of generativity and development. "That is, can such a symbol system develop cognitively on its own, or will such development always be under the auspices of a human programmer?" (p. 159). In characterizing the computer metaphor underlying information processing models of cognitive development, Brown (1979) suggests that:

> A system that cannot grow or show adaptive modification to a changing environment is a strange metaphor for human thought processes, which are constantly changing over the lifespan of the individual and the sociocultural evolution of the race. (p. 227)

De Gelder (1986) maintains that information processing models have neglected what she characterizes as the dual problems of "knowledge dynamics," specifically, that such models fail to adequately address the issues of knowledge change and knowledge use. In this regard she quotes Dennett who suggests that "AI agent creators can in principle make an "adult" agent that is equipped with worldly knowledge as if it had laboriously learned all the things it needs to know" (Dennett, 1984, p. 135). De Gelder (1986) argues that information processing models "idealize away" both the change and use problems. Adult development is treated as something that is given—not to be explained. And the processes by which an individual's "hard-wired" knowledge are to be related to action (i.e., purposive behavior) are ignored.

Siegler, himself a developmentalist strongly identified with information processing theory, asserts: "those of us who have adopted the information process approach have evaded, avoided, and hidden from the issue of the mechanisms by which development occurs" (1984, p. 193).

Klahr (1984) suggests that self-modifying heuristics may be incorporated within production systems to model the processes governing the transitions between states (i.e., development). He points out that production systems as originally conceived were used by developmental psychologists to model state descriptions. But the modularity of production systems makes them well suited for modeling developmental change as well. Basically, this is accomplished by using production systems to modify other production systems. Klahr maintains that such second-order productions may well serve as "the building blocks" for an information-processing theory of developmental transitions. Siegler (1983) is more cautious in his appraisal, questioning whether production systems ought to be viewed as a theory of cognition or a language for describing cognition.

With respect to issues of cognitive development, the information processing approach would seem to offer not so much a comprehensive and generative theory, but rather a powerful methodological tool for testing developmental assumptions once they have been articulated. It is the explicitness and clarity of thinking required in conceptualizing the steps needed to model or simulate a developmental process that is edifying and not the computational metaphor per se.

Information processing on ITCS. As presently articulated, information processing approaches to cognitive development would seem to have little to offer in our understanding of the intergenerational transfer of cognitive skills. While such approaches provide a rich analysis of what it is that develops (i.e., knowledge states), they are at the same time mute with respect to the mechanisms by which these states are acquired and developed over time. Moreover, to argue that an understanding of the logical structure of what is known necessarily explains the underlying psychological processes is to commit the formalistic fallacy, to assume one can necessarily infer function from form.

As suggested earlier, the concept of ITCS by its very nature is a social concept, addressing both what and how cognitive skills are transferred from the members of one generation to those of another. And yet the computer metaphor, and indeed information processing research, has demonstrated a distinctively asocial, individualistic bias. To the extent that mechanisms of learning or development are addressed, they are mechanisms that are to be located exclusively within the individual.

To look for a theory of cognitive development in information processing that will inform our understanding of ITCS at this time is to create a straw man. And yet the preeminence of information processing in current cognitive theory requires that we give such approaches their due. In characterizing information processing approaches to modeling human cognition, Winograd and Flores (1986) suggest that, "At times, justification for the theory seems to be little more than the hope that if the computational system is complex enough and its behavior sufficiently unpredictable, shining intelligence will emerge." Whether powerful computational models of cognitive development will eventually emerge remains to be seen. For now it is suggested that we look elsewhere.

Vygotsky's Sociohistorical Approach

It is tempting to argue that Vygotsky is a developmental theorist of more historical than contemporary import. The fact that he died at the age of 38 of tuberculosis prompted the philosopher, Stephen Toulmin, to liken Vygotsky to the "Mozart of psychology," so active, and yet so short was his life (1978). Bruner (1986) has characterized Vygotsky, along with Freud and Piaget, as one of the "three titans of psychology." Yet, despite Vygotsky's historical status, White (1986) was led to claim "and now Vygotsky has become a theoretical force. It is odd that a man who died in 1934 should have rising fame and influence in 1986." In part, this growing influence can be attributed to the fact that the work of Vygotsky and his colleagues was suppressed within and outside of the Soviet Union from the time that Stalin came to power in the 1930s to his death in 1953. Perhaps most important, however, is the fact that Vygotsky's most distinguished students, Luria and Leont'ev, lived and were active proponents of his sociohistorical position well into the 1970s. Finally, the ideas of Vygotsky are quite in keeping with two closely related theoretical movements that have been growing in importance in contemporary psychology—social constructivism and contextualism. There are a number of scholars of contemporary import whose work has been either informed by or is in the spirit of his ideas. These include Bruner, Cole, Wertsch, Cribner, Harre, Shotter, Gergen, Sampson, and philosophers such as Wittgenstein and Toulmin.

Vygotsky's theory was a much broader conception of development than that of other developmentalists. He believed that developmental analysis was central to psychology as a discipline not just to developmental psychology. In order to understand both what it is that develops and how development occurs, it is essential to understand several important distinctions that he made between different levels of this developmental analysis. These levels include the study of: (a) the evolution of humankind as a species (i.e., phylogenesis); (b) human social history; (c) the development of the individual (i.e., ontogenesis); and (d) development over very short periods of time (microgenesis). In a sense, these levels of analysis may be likened in a series of different lenses that enable us to see varying gradiations between "the wide angle" of development and its finely textured nature.

In particular, Vygotsky believed that if we are to understand both what develops and the means by which this occurs, we must move beyond the individual as the unit of analysis. His was a thoroughgoing social theory of mind. What develops (i.e., ways of knowing and knowledge) are fundamentally social constructions. And how this development occurs is essentially a socially mediated process.

> In order to explain the highly complex forms of human consciousness one must go beyond the human organism. One must seek the origins of conscious activity and "categorical" behavior not in the recesses of the human brain or in the depths of the spirit, but of the external conditions of life. Above all, this means that one must seek these origins in external processes of social life, in the social and historical forms of human existence. (Luria, 1982, p. 25)

What develops? Vygotsky believed that human thinking was largely the result of our ability to appropriate and use culturally evolved tools. The individual's mastery over the tools provided by one's culture plays an essential role in an individual's mastery over the environment and ultimately oneself. The nature of the tools and the manner in which they are acquired and practiced determine what develops. Vygotsky differentiated between material and psychological tools, the former allowing control over environment, the latter over oneself. The former refers to technological artifacts such as the hammer, refrigerator, or ball-bearing; while examples of the latter are represented by sign systems such as those associated with language, mathematics, or art. It was the latter psychological tools, specifically language, that Vygotsky emphasized in the development of thought. He believed that it was through the communicative use of language that individuals are eventually able to regulate their own behavior.

Central to Vygotsky's conception of what develops is the distinction that he made between elementary and higher psychological functions. Elementary mental functions refer to those processes that are characteristic of humans as a species, that are environmentally as opposed to individually regulated, and that are carried out unconsciously. Thus, attention, perception, and memory are all examples of functions that are a part of an individual's natural or biological line of development which may be performed involuntarily. In contrast, higher psychological functions are represented by those conscious processes that have come under individual control. Thus, these uniquely human functions are fundamentally social in nature in two important senses. First, the nature of the sign systems themselves have a sociocultural history. And second, it is through the socially interactive use of signs—most notably language—that individuals are eventually able to engage in voluntary attention or memory. It is not as though the functional use of language merely facilitates already existing processes, but rather, that language makes possible the transformation of elementary processes into functional systems of higher-order processes that would not otherwise occur. Human consciousness emerges with the acquisition and use of language.

The importance attached to the social, cultural, and historical by Vygotsky should be clear. If it is the acquisition of language that so radically transforms the nature of thinking, then the particular language one has acquired and the ways that one has come to use it in practice may vary cross-culturally, and change historically.

If the cognitive processes by which individuals are able to think are fundamentally sociohistorical in nature, so too are the categories with which these processes are associated.

> The system "child-thing" is in reality the system "child-social object." Socially evolved modes of action with these are not given immediately as physical properties of objects. We do not find inscribed on the object where and how it originated. how we may operate with it, how we can reproduce it. Therefore, that object cannot be mastered through adaptation, through a mere "accommodation" to its physical properties. This must take place internally; the child must go through a special process of learning the social modes of action with objects. In this process the physical properties of an object serve merely as referents for the child orientation in his actions with that object. (El'konin, 1972, pp. 237-238)

Important to his conception of development was the distinction Vygotsky made between spontaneous (i.e., everyday) and scientific concepts. The two types of concepts differ along several important dimensions. In general, spontaneous concepts are acquired as a result of direct of unmediated perceptual experience through little or no verbal instruction and consequently with little cultural-historic input. They result in more or less unconscious and episodic forms of behavior. Scientific concepts, in contrast, are acquired as a result of considerably verbally mediated perceptual experience. They are saturated with cultural-historic meanings and are based upon criteria that are typically hierarchically organized and realized consciously by the individual. Thus, one's understanding of the concept of a dog may be either spontaneous or scientific depending upon the extent to which it is informed by cultural-historical derived distinctions (e.g., it is a member of a class of objects that share certain properties, that may in turn be related to other members of its species, which are a part of a larger phylum, etc.). These distinctions may only be realized through verbally mediated perceptual and conceptual experience.

Because a Vygotskian perspective human thinking and its development is so closely dependent upon the nature of and practice with material and psychological tools, development is assumed contingent and open-ended, rather than universal and closed. What is developed through cognitive development, itself, undergoes a process of change over time. Moreover, since the appropriation, internalization, and use of signs and tools is something that may happen throughout one's life, Vygotsky's is a framework that at least enables the construction of a life-span cognitive developmental psychology.

How does development occur? Vygotsky maintained that the mechanisms by which we acquire the ability to think (i.e., elementary processes are transformed into higher psychological processes) are to be found in the social relationship between individuals. In his general law of cultural development, he maintained that all higher psychological functions exist first between individuals and that it is only after extended interactions that individuals are able to internalize and carry them out by themselves.

Given the importance that he attached to the use of signs, it is not surprising that it was in discourse, the verbal interchanges that take place between individuals, where Vygotsky assumed that we would locate the early origins of thinking. However, it wasn't simply the use of signs, but their use within the context of action (i.e., purposive behavior), that Vygotsky assumed was critical for the development of thinking skills.

Opportunities for social interaction leading to development take place within what he termed the zone of proximal development (ZPD), that area of interaction where, unable to solve problems by themselves, individuals are nevertheless able to do so with the assistance of others. It is within these ZPD's that Vygotsky believed that important transformations from other—to self-regulation take place. Essential to this process of internalization is the establishment of intersubjectivity. Intersubjectivity exists when coparticipants come to share the same definition of the task. Thus, while the assumption of an active organism would seem to presuppose an inherent motive-to-learn on the part of the individual, such a motive cannot be assumed to be unbridled. We have seen that what it is that develops is largely channeled by one's culture. The goals and operations used in the service of these goals must be appreciated and appropriated by the individual.

Wertsch (1979) suggest a number of levels of task-related interaction that characterize the transformation from other- to self-regulation. At first, the individual may understand little of the definition or meaning of the task. Gradually the individual comes to understand the definition of the task and assumes increasingly greater strategic responsibility in performing the task. Eventually the individual is able to act independently of the individual who has mediated his or her thinking. Thus, in a seemingly paradoxical way, the individual comes to perform culturally defined tasks well before she or he understands them.

Lest one think that this characterization of the development of higher psychological processes applies exclusively to young children, the reader is reminded of Plato's dialogue between Socrates and Meno concerning whether virtue can be taught. Meno asks, "But how will you look for something when you don't in the least know what it is?" In a similar vein, Schon (1987) has maintained that there is a paradox in an individual's learning a profession such as design in that " 'looking for something' implies a capacity to recognize the thing one looks for, but the student lacks at first the capacity to recognize the object of her search" (Schon, 1987, p. 83).

This is where one may locate the motivational dynamic in Vygotsky's theory. The way in which the knowledgeable individual chooses and frames the task to be learned determine the continued and increasing engagement in the task by the less knowledgeable individual until the learner has both the skill and will to do so herself.

The relationship between intergenerational transfer and cognitive development. There would appear to be considerable overlap between the content and mechanisms of cognitive development with those of intergenerational transfer. From a Vygotskian perspective, both what develops and how this development takes place are closely coupled and are fundamentally social in their origin. Thus, while what is acquired (i.e., higher psychological functions) is closely related to the nature of the tools, both technical and psychological, developed by one's culture, it is the specific practices and the contexts in which these tools have been appropriated and used that determine the nature of mind.

Toward a Developmental Theory of Implicit Conceptions of Cognitive Development

I have concentrated in three theoretical formulations of cognitive development. In addition, there is a fourth conception of cognitive development which might significantly enhance what we know about the intergenerational transfer of cognitive skills—or the lack thereof. I speak of our common sense and often implicit conceptions concerning cognitive development. Fletcher (1984) has differentiated between three important aspects of common sense. They are common sense as: (a) shared fundamental assumptions about the nature of the social and physical world, (b) a set of cultural maxims and shared beliefs about the social and physical world, and (c) a shared way of thinking about the social and physical world (p. 104). The idea of studying common sense, implicit, or naive psychology is not new. However, the idea of investigating our common-sense psychology of cognitive development, and doing so developmentally, has just begun to receive attention. In other words, what are the origins of our common-sense conceptions of the nature of mind and its development, and how do these conceptions change, in turn, as we ourselves develop? In studies illustrative of such an orientation, Lortie (1966) has found that beginning teachers conceptions of teaching and learning are based upon their own "apprenticeships of observation" served when they were students; Goodnow and her colleagues have studied both children and adults' everyday conceptions of learning, development, and intelligence (Goodnow, 1981); Dweck and Bempenchat (1983) have investigated children's theories of intelligence, and Sternberg and Berg (1987) have studied people's implicit theories of intelligence at different points in adult intellectual development. Not only is it the case that our cognitive development is socially mediated, but, so too, are the ways that we understand this development in ourselves and others.

If, as suggested by Thomas (1928), situations defined as real are real in their consequences, then our beliefs about the nature of cognitive development may influence to a considerable extent what we do, or don't do, to see that such development is promoted. Because individuals' beliefs are often implicit, it is necessary that we examine their assumptive frameworks as revealed in practice as well as those common-sense conceptions that they are able to verbalize. Indeed, much that is implicit concerning cognition and its development may carry ideological implications in that it is grounded in the thinking and reasoning manifest in the practices, tasks, and activities of the dominant cultural view of our society (Sampson, 1981).

Cognitive Development as Ideology. In an article that seems to have been ignored by most cognitive and developmental psychologists, Sampson (1981) urges that we consider the ideology implicit in cognitive and developmental psychologies. He defines ideology as a set of values, interests, and social practices that reproduce the existing nature of the social order. The very fact that we are examining the *intergenerational* transfer of cognitive skills in this conference would seem to entail an ideological presupposition that mind is a social and not simply an individual construction.

Consider the ideological implications for how we think about the relationship between ITCS and cognitive development that inheres in the values, interests, and social practices of the following instances.

- The fact that we have become an increasingly age-differentiated and age-segregated society.
- Gerald Coles' (1987) compelling argument for the notion that the learning disability label is given selectively to middle-class children while the learning difficulties of children from minority and poor communities are explained to be the result of emotional or intellectual problems.
- Research by Linda Anderson (1985) that indicates that upwards of 70 percent of elementary school children's time is spent working alone at their seats.
- A recent segment on NPR's "All Things Considered" detailing the resistance of parent groups to the introduction of collaborative learning experiences in their children's classrooms. The argument is that such programs are leftist-inspired.

Sampson argues that in addition to the individualistic bias as exemplified above, current conceptions of cognitive development privilege certain forms of knowledge over others. Thus, from a Piagetian perspective, scientific rationality is taken as the model toward which we all develop.

Sampson criticizes much of contemporary cognitive psychology for the strong individualistic and subjectivistic biases that it has demonstrated in its theory and research. Not surprisingly, many of our ideological beliefs in psychology mirror those found in the public at large. Psychology not only reflects, but at the same time further extends these beliefs in the goals and practices that it adopts. The relationship is mutual and dialectic.

An understanding of the development of our common-sense beliefs concerning cognition and its development along with the implicit ideological underpinnings of these beliefs carries both theoretical and practical implications. Theoretically, we have seen that there is compelling evidence to suggest that both what development is, and how it occurs, are culturally and historically contingent. Moreover, an understanding of the received view of cognitive development held by the various publics with which we interact and whom we serve carries with it important policy and practical implications.

REFERENCES

Anderson, L. M., Brubaker, N. L., Alleman-Brooks, J., & Duffy, G. (1985). A qualitative study of seatwork in first-grade classrooms. *The Elementary School Journal, 86*, 1-20.

Baltes, P. B., & Goulet, L. R. (1970). Status and issues of a life-span developmental psychology. In L. R. Goulet & P. B. Baltes (Eds.), *Life-span developmental psychology: Dialectical perspectives*. New York: Academic Press.

Brown, A. (1979). Theories of memory and the problems of development: Activity, growth and knowledge. In L. S. Cermak & F. I. M. Craik (Eds.), *Levels of processing in human memory* (pp. 225-258). Hillsdale, NJ: Erlbaum.

Bruner, J. (1986). *Actual minds, possible worlds*. Cambridge, MA.: Harvard University Press.

Cole, M. (Speaker). (1988). *A cultural approach to learning and development: Implications for educational research and practice* (Cassette recording no. RA8-5.06). New Orleans, LA: American Educational Research Association.

Coles, G. (1987). *The learning mystique: A critical look at "learning disabilities."* New York: Pantheon Books.

De Gelder, B. (1986). On what is cognitive about development. In P. L. C. Geert (Ed.), *Theory building in developmental psychology*. North-Holland: Elsevier Science Publishers.

Dennett, D. C. (1984). Cognitive wheels: The frame problem of AI. In C. Hookway (Ed.), *Minds, machines and evolution*. Cambridge, UK: Cambridge University Press.

Dweck, C. S., & Bempechat, J. (1983). Children's theories of intelligence: Consequences for learning. In S. G. Paris, G. M. Olson, & H. W. Stevenson (Ed.), *Learning and motivation in the classroom* (pp. 239-256). Hillsdale, NJ: Erlbaum.

Edwards, D., & Middleton, D. (1986). Conversation with Bartlett. *The Quarterly Newsletter of the Laboratory of Comparative Human Cognition, 8*, 79-89.

El'konin, D. B. (1971). Toward the problem of stages in the mental development of the child. *Soviet Psychology, 10*, 225-251.

Fletcher, G. G. O. (1984). Psychology and commonsense. *American Psychologist, 39*, 203-213.

Gavelek, J. R. (1986). The social contexts of literacy and schooling: A developmental perspective. In T. E. Raphael (Ed.), *The contexts of school-based literacy* (pp. 3-26). New York: Random House.

Gavelek, J. R. (in press). Social Constructivism as an alternative conception of reading (dis)ability. *International Journal of Educational Research.*

Gergen, K. J. (1980). The emerging crisis in life-span developmental theory. In P. B. Baltes & O. G. Brim, Jr. (Eds.), *Life-span development and behavior* (Vol. 3). New York: Academic Press.

Gergen, K. J. (1985). The social constructionist movement in modern psychology. *American Psychologist, 40*, 266-275.

Goodnow, J. (1981). Everyday ideas about cognitive development. In J. P. Forgas (Ed.), *Social cognition: Perspectives on everyday understanding* (pp. 85-112). New York: Academic Press.

Harre, R. (1986). Social sources of mental content and order. In J. Margolis, P. T. Manicas, R. Harre, & P. F. Secord (Ed.), *Psychology: Designing the discipline* (pp. 91-127). Oxford: Basil Blackwell.

Kessen, W. (1984). Introduction: The end of the age of development. In R. J. Sternberg (Ed.), *Mechanisms of cognitive development* (pp. 1-17). New York: Freeman.

Klahr, D. (1984). Transition processes in quantitative development. In R. J. Sternberg (Ed.), *Mechanisms of cognitive development* (pp. 106-139). New York: Freeman.

Langley, P. (1983). Exploring the space of cognitive architectures. *Behavioral Research Methods and Instrumentation, 15*, 289-299.

Lortie, D. C. (1975). *Schoolteacher.* Chicago: University of Chicago Press.

Luria, A. R. (1982). *Language and cognition.* New York: Wiley & Sons.

Mehan, H. (1981). Social constructivism in psychology and sociology. *The Quarterly Newsletter of the Laboratory of Comparative Human Cognition, 3*, 71-77.

Neisser, U. (1976). *Cognition and reality.* San Francisco: Freeman.

Newell, A. (1973). Production systems: Models of control structures. In W. G. Chase (Ed.), *Visual information processing.* New York: Academic Press.

Norman, D. A. (1981). Twelve issues for cognitive science. In D. Norman (Ed.), *Perspectives on cognitive science* (pp. 265-295). Norwood, NJ: Ablex.

Piaget, J. (1954). *The child's construction of reality*. New York: Basic Books.

Piaget, J. (1964). *Learning and development.*

Piaget, J. (1984). Piaget's theory. In P. H. Mussen & W. Kessen (Eds.), *Handbook of child psychology* (Vol. 1, pp. 103-128). New York: Wiley & Sons.

Rabinowitz, F. M., Grant, M. J., & Dingley, H. L. (1987). Computer simulation, cognition, and development: An introduction. In J. Bisanz, C. J. Brainerd, & R. Kail (Eds.), *Formal methods in developmental psychology* (pp. 263-301). New York: Springer-Verlag.

Russell, J. (1978). *The acquisition of knowledge*. New York: St. Martin's Press.

Sampson, E. E. (1981). Cognitive psychology as ideology. *American Psychologist, 36*, 730-743.

Schon, D. A. (1987). *Educating the reflective practitioner*. San Francisco: Jossey-Bass.

Shaw, R., & Bransford, J. (1979). Introduction: Psychological approaches to the problem of knowledge. In R. Shaw & J. Bransford (Ed.), *Perceiving, acting and knowing*. New York: John Wiley & Sons.

Siegler, R. S. (1984). Information processing approaches to development. In P. H. Mussen & W. Kessen (Eds.), *Handbook of child psychology* (Vol. 1, pp. 129-212). New York: Wiley & Sons.

Simon, H. A. (1980). Cognitive science: The newest science of the artificial. *Cognitive Science, 4*, 33-46.

Sternberg, R. J., & Powell, J. S. (1984). The development of intelligence. In P. H. Mussen & W. Kessen (Eds.), *Handbook of child psychology* (Vol. 3, pp. 341-419). New York: Wiley & Sons.

Sternberg, R. J., & Berg, C. A. (1987). What are the theories of adult intellectual development theories of? In C. Schooler & K. W. Schaie (Eds.), *Cognitive functioning and social structure over the life course* (pp. 3-23). Norwood, NJ: Ablex.

Teske, J. A., & Pea, R. D. (1981). Metatheoretical issues in cognitive science. *The Journal of Mind and Behavior, 2*(2), 123-178.

Thomas, W. I. (1928/1966). Situational Analysis: The behavior pattern and situation. In R. M. Janowitz (Ed.), *W. I. Thomas on Social Organization and Social Personality* (pp. 23-41). Chicago: University of Chicago Press.

Toulmin, S. (1978). The Mozart of psychology. New York *Review of Books*, pp. 51-57.

Toulmin, S. (1979). The inwardness of mental life. *Critical Inquiry, 6*, 1-16.

Toulmin, S. (1981). Epistemology and developmental psychology In E. S. Gollin (Ed.), *Developmental plasticity* (pp. 253-267). New York: Academic Press.

Vygotsky, L. S. (1981). The genesis of higher mental functions In J. V. Wertsch (Ed.), *The concept of activity in Soviet psychology*. Armonk, NY: M. E. Sharpe.

Vygotsky, L. S. (1986). *Thought and language* (A. Kozulin, Ed. & Trans.). Cambridge, MA: MIT Press. (Original work published 1934)

Wertsch, J. V. (1979). From social interaction to higher psychological processes: A clarification and application of Vygotsky's theory. *Human Development, 22*, 1-22.

Wertsch, J. V. (1985). *Vygotsky and the social formation of mind.* Cambridge, MA: Harvard.

White, S. H. (1986). Cited in L. S. Vygotsky *Thought and language.* Cambridge, MA: The MIT Press.

Winograd, T., & Flores, F. (1986). *Understanding computers and cognition.* Norwood, NJ: Ablex Publishing.

Chapter 3

The Intergenerational Transfer of Motivational Skills

Janine Bempechat
Harvard University

The focus of the work presented in this edited volume is on the intergenerational transfer of cognitive skills. In this chapter, I will turn the focus onto the motivational aspects of learning and argue that our understanding of the socialization of cognitive skills is incomplete without an understanding of the socialization of motivational skills.

COGNITIVE SKILLS FROM THE PERSPECTIVE OF MOTIVATION THEORY

In theory, children can arrive at school armed with all the cognitive skills they need to succeed. They will not reach their potential, however, if they are prone to maladaptive motivational tendencies, such as a lack of persistence, a preference for easy over challenging tasks, low expectations for success, attributions to ability or lack of ability, or a propensity to quit in the face of difficulty.

Why are motivational factors important? They are important because there is evidence suggestive that children's performance in school is predicted more reliably from motivational factors than from actual measures of intelligence (Crandall, 1969; Dweck & Licht, 1980; Weiner, 1972). This implies that there are very bright children who are vulnerable to learned helplessness and who consistently avoid challenge. This also implies that there are children who are not as "bright" but who thrive on challenge and may attain levels of proficiency that we would never have predicted from knowing their intelligence or achievement scores. Why might this happen?

The work of Dweck (Dweck & Bempechat, 1983) and Nicholls (Jagacinski & Nicholls, 1984; Nicholls, 1978, 1983, 1984a,b) suggests that children's beliefs about the nature of intelligence may influence their achievement cognitions and behaviors. These researchers and their colleagues have respectively categorized children's beliefs about ability as either fixed and static *(entity* or *differentiated)* or as malleable and flexible *(incremental* or *undifferentiated)*. They have also shown that children who think that intelligence is fixed tend to choose tasks which display their ability ("performance" goals/ego involvement) rather than challenge it ("learning" goals/task involvement) (Bandura & Dweck, 1985). These children tend to be vulnerable to learned helplessness, particularly when their confidence is low (Elliott & Dweck, 1985) and also tend to define intelligence in others in terms of static qualities ("She's smart because she always gets A's") (Bempechat, London, & Dweck, in press).

In contrast, children who view intelligence as a fluid, changeable quality tend more to undertake challenging tasks, even at the risk of failure (Bandura & Dweck, 1985). They also tend to exhibit more mastery-oriented behavior regardless of their confidence level and to be resilient in the face of failure (Elliott & Dweck, 1986). In addition, they tend to define intelligence in others on the basis of actions (e.g., "He's smart because he always does his homework") (Bempechat et al., in press) rather than of static qualities. These differences in achievement orientation are independent of actual skill or measured intelligence.

A recent study found that children's beliefs about intelligence can be successfully manipulated by verbal instructions (Bempechat et al., in press). This implies that children can be oriented towards a more flexible view of ability, which may be more adaptive in the long run.

We do know that beliefs about achievement can be influenced by classroom environment. Ames and Ames have shown that competitive learning environments inhibit effort-related cognitions and make self-perceptions of ability salient (Ames, Ames, & Felker, 1977; Ames & Felker, 1979). Individualistic learning environments tend to foster effort attributions and adaptive strategies, such as self-monitoring and self-instructions (Ames, 1984; Ames & Ames, 1981).

However, we also know that there are individual differences in beliefs about achievement and achievement behaviors. How do these individual differences arise? More specifically, what role does the family play in fostering achievement beliefs and behaviors?

THE ROLE OF PARENTAL BELIEFS

The burgeoning research on parental beliefs is converging to show that parents' own achievement beliefs and attitudes towards education serve to organize and guide their behavior with their children (Eccles, 1983; Entwisle, Alexander, Pallas, & Cadigan, 1987; Entwisle & Hayduk, 1988; McGillicuddy-DeLisi, 1985; Miller, 1986, 1988; Phillips, 1987; Sigel, 1985; Thomas, in Phillips, 1987). And, as Phillips (1987) has argued, parents' beliefs exert a strong and perhaps causal influence on their children's developing achievement attitudes and behaviors. Nowhere is this more striking than in the research on sex differences in math achievement. Recent research has shown that parents' beliefs have a profound influence on children's self-appraisals of ability, attributions for performance, and attitudes towards math.

In a recent study, Hess and his colleagues (Holloway & Hess, 1985; Dunton, McDevitt, & Hess, 1988) found that, even though there were no significant sex differences on several assessments of performance, mothers expressed different attributions for girls' and boys' achievement. In explaining high achievement, mothers of boys place more emphasis on ability than effort, while mothers of girls place more emphasis on effort than ability. In explaining low achievement, mothers of boys stress lack of effort over lack of ability, but express the opposite attribution for their girls.

It is not surprising, then, to learn that compared to boys, girls have lower self-concept of math ability. Girls also tend to believe that math is harder and of less general value (Eccles, 1983; Parsons, Adler, & Kaczala, 1982). Moreover, they are less likely to attribute success to ability and more likely to attribute it to stable effort. They are also more likely to attribute failure to lack of ability. These researchers have shown that children's self-perceptions of math ability appear to be influenced more by their parents' appraisals than by their own record of achievement.

In a longitudinal study that followed children from pre-kindergarten through tenth grade, Stevenson and Newman (1986) found that sex differences in self-ratings of general and specific cognitive abilities emerged only in tenth grade. At this point, boys had more positive attitudes towards math than girls. In addition, boys had higher ratings than girls on expectancies for success in a career that required math ability, the utility of math for daily life, and ease of math courses.

These sex differences in attitudes were not accounted for by performance differences. For boys, their math performance in third and fifth grades were the only variables related to their later attitudes. In contrast, girls' later attitudes were accounted for by previous performance and by mother and teacher ratings of their cognitive skills. Thus, mothers' perceptions appear to have an important influence on their children's achievement-related attitudes, especially those of their girls.

As reported elsewhere in this volume (see Ginsburg, Bempechat, & Chung), the beliefs of Asian parents regarding the malleability of ability appear to have an important effect on their children's performance in school. Academic achievement in Asian cultures, therefore, provides an interesting context in which to examine some principals of motivation theory. This issue is examined below.

MOTIVATION THEORY APPLIED TO ACADEMIC ACHIEVEMENT IN ASIAN CHILDREN

In our own work on the school achievement of Asian American children, Herb Ginsburg and I (see this volume) are examining the degree to which immigrant status and a common Confucian ethic may influence the socialization practices of Asian American parents, such that they foster motivational and attitudinal tendencies that contribute to high achievement. In conceptualizing the research issues, we are relying heavily on the attribution theory of Bernard Weiner. For example, Weiner and his colleagues (Weiner, 1972, 1980, 1984) have shown that attributions to internal, controllable factors, such as effort, are very adaptive because they propel individuals towards positive action (i.e., studying). The Asian cultural adherence to the value of personal effort in achievement, coupled with the minimal value placed on innate ability, may serve to foster a sense of control and personal responsibility towards school learning.

This sense of control/responsibility over one's learning is associated with mastery-oriented behavior. That is, children who take responsibility for their own academic successes and failures are more likely to confront difficulty or challenge with renewed resolve, and to engage in adaptive methods of problem solution (i.e., self-monitoring, self-instruction). In contrast, children who blame other people or external events for their school performance are more likely to succumb to learned helplessness in the face of difficulty, and to engage in ineffectual problem-solving strategies (Diener & Dweck, 1978, 1980).

In addition to the value that parents place on effort, some research suggests that parents' reactions to failure may play an equally important role in maintaining high effort. According to Weiner's work on attribution-affect linkages (Weiner, Graham, Stern, & Lawson, 1982), the affective reactions of others are likely to be important determinants of self-ascriptions for success and failure. Interestingly, in a comparative study of parents' beliefs about children's performance in mathematics, Hess, Chih-Mei, and McDevitt (1987) found that, while Chinese American and Caucasian parents tried to determine the reasons for their children's failure, mothers in the People's Republic of China punished their children and were angry with them when they failed. This finding is interesting in light of Weiner et al.'s (1982) study. They showed that, in a hypothetical scenario, when teachers react to a student's failure with anger, children infer that the cause of this failure is lack of effort. It is possible that the anger of Chinese mothers in response to their children's failure would serve to communicate to the children that their mothers fully expect them to try harder, and in so doing, they will perform better. The effect on the children could be to foster renewed high effort.

With respect to parental expectancies, the literature suggests that Asian parents expect more of their children than do American parents, and hold them up to higher standards. Recent research has shown that, despite the higher achievement of their children, Asian parents are less satisfied with their children's performance than are American parents, adhere to higher standards in judging their academic record, and have higher expectations for their performance (Stevenson, Lee, & Stigler, 1986; Uttal, Chen & Diaz-Perez, 1987; see Ginsburg et al., this volume, for a more detailed treatment of this issue). Of course, research on teacher expectancies has shown that high expectancies do appear to have a positive influence on students' performance (Rosenthal, 1985; Rosenthal & Jacobson, 1968).

While some controversy over this research still exists, it is important to bear in mind that, in Rosenthal's original research, he led teachers to believe that some children would "bloom" intellectually over the coming year (see Dweck, 1988, for a further discussion of this issue). The implicit message was that intelligence develops and is malleable. Given that this belief is very strongly held in Asian cultures, one is led to suggest that the relatively high expectancies of Asian parents may communicate an implicit belief that there are no limits on intellectual growth. As current research is showing, this belief appears to be a strong motivator in children's learning.

With these research findings in mind, one hypothesis that we are exploring is the possibility that Asian American children may be socialized to be more resilient to learned helplessness than their Caucasian peers. If these children are taught from an early age that their academic future lies essentially in their own hands, passivity in the face of school difficulty may simply not be perceived as a viable option for them. A strong belief in the value of effort may indeed foster a strong sense of discipline that might help these children persevere during difficult and challenging moments in their education.

SUMMARY

In sum, I believe that the question before researchers and educators should be two-fold: (a) how do parents transmit to their children the cognitive skills necessary to succeed in school, and (b) how do parents transmit adaptive motivational tendencies that can maximize their children's cognitive skills? A research agenda that approaches both issues simultaneously can be very successful in identifying a set of child-rearing practices that enhance both cognitive skills and adaptive achievement cognitions and behavior. With such practices in hand, we can ultimately develop intervention strategies for parents that will train them to foster the development of their children's cognitive skills and teach their children how to make the most of these skills throughout their school years.

REFERENCES

Ames, C. (1984). Achievement attributions and self-instructions under competitive and individualistic goal structures. *Journal of Educational Psychology, 76,* 478–487.

Ames, C., & Ames R. (1981). Competitive versus individualistic goal structures: The salience of past performance information for causal attribution and affect. *Journal of Educational Psychology, 73,* 411–418.

Ames, C., Ames, R., & Felker, D. (1977). Effects of competitive reward structure and valence of outcome on children's achievement attributions. *Journal of Educational Psychology, 69,* 1–8.

Ames, C., & Felker, D. (1979). An examination of children's attributions and achievement-related evaluations in competitive, cooperative, and individualistic reward structures. *Journal of Educational Psychology, 71,* 413–420.

Bandura, M. M., & Dweck, C. S. (1985). *Self-conceptions and motivation: Conceptions of intelligence, choice of achievement goals, and patterns of cognitions, affect, and behavior.* Unpublished manuscript.

Bempechat, J., London, P., & Dweck, C. (in press). Children's conceptions of ability in major domains: An interview and experimental study. *Child Study Journal.*

Crandall, V. C. (1969). Sex differences in expectancy of intellectual and academic reinforcement. In C. P. Smith (Ed.), *Achievement-related motives in children.* New York: Russell Sage.

Diener, C., & Dweck, C. S. (1978). An analysis of learned helplessness: Continuous changes in performance, strategy, and achievement cognitions following failure. *Journal of Personality and Social Psychology, 36,* 451–462.

Diener, C., & Dweck, C. S. (1980). An analysis of learned helplessness II. The processing of success. *Journal of Personality and Social Psychology, 39,* 940–952.

Dunton, K., McDevitt, T., & Hess, R. (1988). Origins of mothers' attributions about their daughters' and sons' performance in mathematics in sixth grade. *Merrill-Palmer Quarterly, 34,* 47–70.

Dweck, C., & Bempechat, J. (1983). Children's theories of intelligence: Consequences for learning. In S. Paris, G. Olson, & H. Stevenson (Eds.), *Learning and motivation in the classroom.* Hillsdale, NJ: Erlbaum.

Dweck, C. S., & Licht, B. G. (1980). Learned helplessness and intellectual achievement. In J. Garber & M. E. P. Seligman (Eds.), *Human helplessness: Theory and applications.* New York: Academic Press.

Dweck, C. S. (1988). Maturation. In R. Glaser & A. Lesgold (Eds.), *The handbook of psychology and education.* Hillsdale, NJ: Erlbaum.

Eccles, J. (1983). Expectancies, values, and academic behaviors. In J. Spence (Ed.), *Achievement and achievement motives: Psychological and social approaches.* New York: Freeman.

Elliott, E., & Dweck, C. S. (1985). *Goals: An approach to motivation and achievement.* Unpublished manuscript.

Entwisle, D., & Hayduk, (1988). Lasting effects of elementary school. *Sociology of Education, 61,* 147–159.

Entwisle, D., Alexander, K., Pallas, A., & Cadigan, D. (1987). The emergent academic self-image of first graders: Its response to social structure. *Child Development, 58,* 1190–1206.

Hess, R., Chih-Mei, C., & McDevitt, T. (1987). Cultural variations in family beliefs about performance in mathematics: Comparisons among People's Republic of China, Chinese-American and Caucasian-American families. *Journal of Educational Psychology, 79,* 179–188.

Holloway, S., & Hess, R. (1985). Mothers' and teachers' attributions about children's mathematics performance. In I. Sigel (Ed.), *Parental belief systems.* Hillsdale, NJ: Erlbaum.

Jagacinski, C. M., & Nicholls, J. G. (1984). Conceptions of ability and related affects in task involvement and ego involvement. *Journal of Educational Psychology, 76,* 909–919.

McGillicuddy-DeLisi, A. (1985). The relationship between parental beliefs and children's cognitive level. In I. Sigel (Ed.), *Parental belief systems.* Hillsdale, NJ: Erlbaum.

Miller, S. (1986). Parents' beliefs about their children's cognitive abilities. *Developmental Psychology, 22,* 276–284.

Miller, S. (1988). Parents beliefs about children's cognitive development. *Child Development, 59,* 259–285.

Nicholls, J. G. (1978). The development of the concepts of effort and ability, perception of own attainment, and the understanding that difficult tasks demand more ability. *Child Development, 49,* 800–814.

Nicholls, J. G. (1983). Conceptions of ability and achievement motivation: A theory and its implications for education. In S. G. Paris, G. M. Olson, & H. W. Stevenson (Eds.), *Learning and motivation in the classroom.* Hillsdale, NJ: Erlbaum.

Nicholls, J. G. (1984a) Conceptions of ability and achievement motivation. In R. Ames & C. Ames (Eds.), *Research on motivation in education: Student motivation.* New York: Academic Press.

Nicholls, J. G. (1984b). Achievement motivation: Conceptions of ability, subjective experience, task choice, and performance. *Psychological Review, 91,* 328–346.

Parsons, J., Adler, T., & Kaczala, C. (1982). Socialization of achievement attitudes and beliefs. *Child Development, 53,* 310–321.

Phillips, D. (1987). Socialization of perceived academic competence among highly competent children. *Child Development, 58,* 1308–1320.

Rosenthal, R. (1985). From unconscious experimenter bias to teacher expectancy effects. In *Teacher expectancy effects.* Hillsdale, NJ: Erlbaum.

Rosenthal, R., & Jacobson, L. (1968). *Pygmalion in the classroom.* New York: Holt, Rinehart & Winston.

Sigel, I. (1985). A conceptual analysis of beliefs. In I. Sigel (Ed.), *Parental belief systems.* Hillsdale, NJ: Erlbaum.

Stevenson, H., Lee, S., & Stigler, J. (1986). Mathematics achievement of Chinese, Japanese, and American children. *Science, 231,* 693–699.

Stevenson, H., & Newman, R. (1986). Long-term prediction of achievement and attitudes in mathematics and reading. *Child Development, 57,* 646–659.

Thomas, N. (1985). *The effects of parental attitudes and classroom climate on children's self-concept of mathematics ability.* Unpublished doctoral dissertation, University of Illinois at Urbana-Champaign.

Uttal, D., Chen, C., & Diaz-Perez, R. (1987, April). *Parental beliefs and children's school achievement in the United States and China.* Paper presented at the biennial meeting of the Society for Research in Child Development, Baltimore.

Weiner, B. (Ed.). (1972). *Theories of motivation: From mechanism to cognition.* Chicago: Markham.

Weiner, B. (1980). May I borrow your class notes? An attributional analysis of help giving in an achievement-related context. *Journal of Educational Psychology, 72,* 676–681.

Weiner, B. (1984). Principles for a theory of student motivation and their application within an attributional framework. In R. Ames & C. Ames (Eds.), *Research on motivation in education: Student motivation.* New York: Academic Press.

Weiner, B., Graham, S., Stern, P., & Lawson, M. (1982). Using affective cues to infer causal thoughts. *Developmental Psychology, 18,* 278–286.

Chapter 4

Literacy Transfer from Parents to Children in the Preschool Years

Jana M. Mason
Bonnie M. Kerr
University of Illinois

According to Stedman and Kaestle (1987), as much as 20 percent of the adult population in the United States, or around 35 million people, have serious difficulties with everyday reading tasks such as a six-word direction on a medicine bottle, a sentence about how often houses should be inspected for termites, airline and train schedules, or a child's report card. Another 10 percent or so are probably marginal in their functional literacy skills. The problem is so serious that "among the lowest fifth in functional literacy skills are many who are unable to read product labels and have to depend upon brand-name logos for selection of items in a grocery store. "[As a result], parents' roles as guardians of their children could be seriously undermined by a lack of functional reading skills. Many cannot read letters or help their children with homework. Some may have trouble finding emergency phone numbers for police and fire stations. Such low literacy skills can be crippling, and a target group of 20% of our population is alarming" (p. 34).

Stedman and Kaestle are not alone in their concern about low adult literacy levels. Carroll (1987) reviewed work from national assessments conducted from 1971 to 1985. He reports that the median literacy performance at age 9 is a little beyond the Basic level (reading by locating simple facts). By age 13 it is a little beyond Intermediate (reading by searching, organizing, and paraphrasing information), and at age 23 it is somewhat better than Adept (reading complicated literary and informational passages). However, about 44 percent of young adults have not yet attained the Adept level. In 1984, nearly 25 percent of the urban poor scored no better than the Basic level and about half scored no higher than the Intermediate level. Moreover, trends over a 13-year period suggest that, except for a reduction in the number of individuals scoring below the Basic level among the urban poor, there has been little change.

Inadequate adult literacy levels in the United States do not stand alone. Affected as well are the literacy levels of the next generation. Because the development of reading and writing knowledge begins long before formal schooling, parental influence is critical. We don't yet understand the mechanisms by which this influence occurs. What dispositions, skills, and abilities with regard to literacy develop during the preschool years? How do parents help their children learn to become literate? Answers to these questions might allow intergenerational cycles of illiteracy to be broken.

COMPONENTS OF LITERACY ACQUIRED DURING THE PRESCHOOL YEARS

We propose that the development of literacy has at least three major components that are tied directly to using written language. They are: phoneme and grapheme awareness, concepts about print, and features of written language. After explaining these components, we suggest how literate parents transfer them to their children. Then we discern the problem confronting parents who may not be literate.

Phonome and Grapheme Awareness

Phonological awareness is an important cognitive mechanism that underlies reading acquisition. Phonological awareness is defined as "conscious access to the phonemic level of language and some ability to cognitively manipulate representations at this level" (Stanovich, 1986, p. 14). It has been shown that phonological awareness measures predict reading ability even when assessed before a child is able to read (Adams, 1990; Lundberg, 1986; Tunmer, Herriman, & Nesdale, 1988). Furthermore, when phonological awareness skills are taught to young children, significant advantages in reading occur (Bradley & Bryant, 1985; Ehri & Wilce, 1987).

Awareness of separate phonemes has been found to be developmental in nature. Treiman (1986) presents the thesis that the development of awareness about phonemes has steps or stages, with the units that are intermediate between syllable and phoneme playing an important early role. Syllables in words are distinguished first. Then the syllable, with two major subunits, called the *onset* and the *rime,* are identified next. The onset is the initial consonantal portion of the syllable (e.g., *m* as in *mask*) and the rime must contain a peak or vowel nucleus (e.g., *i* in *milk*) and may also contain an ending consonant pattern (e.g., *sk* as in *mask*). Children can analyze spoken syllables into onsets and rimes before they can identify phonemes. She explained, "This difference is consistent with the notion that single initial consonants are units on their own, while final consonants are part of a large unit, the rime. It is easier to divide the onset from the rime (which permits success in the initial consonant task) than to divide the rimes into phonemes (which is necessary for the final consonant task)" (p. 14).

Similarly, but focusing on the later stages of the development of phonemic awareness, Goswami and Bryant (1987) hypothesize that not only rhyming, but also generalization, occur before or while children are learning to read. Rhyming involves the recognition of similar ending patterns in words and the ability to break words up into at least two segments. Generalization involves deciphering unknown written words by thinking of similar (rhyming or alliterative) words. This is referred to as decoding by analogy. Studies show that rhyming intervention produces a significant improvement in reading, and kindergarten and primary grade children make analogies as they figure out how to read real words and pseudowords.

Earlier work in phonemic awareness has a long tradition labeled as auditory discrimination (Gates, Bond, & Russell 1939; Monroe, 1932; Durrell & Murphy, 1953). Measured principally with tests to discriminate words that differ by one phoneme, phonemic awareness has been found to be correlated .40 or better with other reading measures and to differentiate good from poor readers (reviewed by Mason, 1984). More recent work continues to show effects with measures that include hearing beginning and ending word sounds, segmenting words into phonemes, and deleting or adding phonemes from words to create other words (Stanovich, Cunningham, & Cramer, 1984; Yopp, 1983).

Graphic awareness is based on a long tradition which assumes that visual recognition and discrimination of letters and words is critical to reading development. A review of 30 years of research by Barrett (1965) links letter-naming knowledge and reading, and research today continues to show correlations above .50. Ehri (1983) and Mason (1984) suggest that letter-name instruction may contribute to learning to read because letter names contain sounds of the letters in words. Hence, letter-name knowledge can foster phonemic awareness and lead to an understanding of phoneme–grapheme awareness (e.g., replacing one initial consonant with another in words, such as replacing *b* in *bat* with *h* to make *hat*). Ehri and Wilce (1987) provide evidence that the construct of phoneme-grapheme correspondence interacts with the learning task, showing that whether children learn words in isolation or in sentences affects their later knowledge about the words. That is, the context for good learning also matters.

CONCEPTS ABOUT PRINT

Although necessary for reading, phoneme–grapheme correspondence is not sufficient. To become a fluent reader, according to Perfetti (1986) and Stanovich (1986), children must also acquire techniques that allow rapid and automatic processing of print. They need to perceive about two words at a time as they read, identify visual patterns quickly, and be able to coordinate processes of reading with recall of information about what is being read. Language comprehension skills are also necessary, particularly an ability to focus on and interpret larger units of information. In addition, reading requires flexibility, an ability to attend to the meaning of information and its properties at the same time, or to "decenter," using the Piagetian term (Tunmer et al., 1988). Children need to sort letters from words, words from referents, and context from grammatical forms in order to maintain a balance between text meaning and component elements (Lartz & Mason, 1988). These additional abilities require the development of a wide-ranging set of concepts about print and an awareness of how to analyze language and view it out of context.

Analyses of reading tests given to 3–6 year-old middle-class children (Lomax & McGee, 1987) document the importance in literacy development of print concepts in conjunction with phonological insights. In this research, a five-component model was determined: concepts about print, graphic awareness, phonemic awareness, grapheme–phoneme correspondence, and word reading. Although overlapping with and affecting both graphic awareness and grapheme–phoneme awareness, the components developed in approximately that order, beginning with concepts about print.

The concepts about print components, based on work by Clay (1979), Downing, Ayers, and Schaefer (1978), and Mason (1980), refers to knowledge about the process of reading. It includes abilities to identify environmental printed labels and signs in context; knowing where to begin and end reading, what direction to read, how to integrate what is read; distinguishing printed letters, words, and sentences; and identifying types of printed materials (e.g., books, newspapers, calendars). In the Lomax and McGee study all of these concepts were partially developed by age 3, and most were fully developed by age 5.

WRITTEN LANGUAGE FEATURES

Written language is not only a way to convey oral information; it has a structure that can be studied, analyzed, and interpreted (Olson, 1984). Several characteristics of written language are important for understanding the interactive nature of parents transferring literacy knowledge during the preschool years. To begin, written language is often removed from the here and now; it is decontextualized (Chafe, 1982, 1985; Perera, 1984; Tannen, 1982). As a result, text meaning must be carried by semantic and syntactic systems and done without the paralinguistic cues or a shared physical context that characterize oral language communication. Torrance and Olson (1985) found that children more skilled in using semantic and syntactic structures of language, particularly cognitive verbs and subordination, were better in beginning reading.

Perera (1984) has a framework for specifying oral and written language differences which include the following categories: *Physical* (differences involving modality and time), *Situational* (differences involving the level of interactive contexts), *Functional* (differences involving propriety of use), *Form* (differences involving format and prosodic and paralinguistic features), and *Structural* (differences involving syntactic and discourse features). Defining the differences between oral and written language in this manner allows for understanding where beginning problems can arise. This explains why Dyson (1984) found that children who have had few literacy experiences become easily frustrated. They are unable to map words onto their oral language because they have not mastered the differences in form and structure.

LITERACY TRANSFER FROM PARENTS TO CHILDREN

Parents promote literacy for children through a wide variety of language and print experiences. Through these experiences parents actually teach phoneme and grapheme correspondences and concepts about print, with distinctions between oral and written language features thereby untangled. Reading stories to children captures print and language experiences, so it may be the most effective way to promote literacy in the preschool years.

Discussion about stories is also important. When reading stories to children, parents interact in ways which allow children to use what they know about book concepts and about the topic of the story while they are learning new concepts. For example, children not yet reading can interpret picture clues and the parent can point to connections between picture and print. Children can make sense of the picture while parents make sense of the text for them. Vygotsky (1978) describes this type of interaction as working within the zone of proximal development. When parents read stories to children, they operate within this zone (Snow, 1983).

An example of operating within the zone of proximal development can be seen in the following excerpt of a mother reading a book called *Snowman* to her 4-year-old daughter, Heather (information excerpted from an ongoing study investigated by Kerr, Mason, & McCormick, 1991).

Mother: Right here. We'll try this one. [The mother puts the book in front of Heather]. What is that? [Mother points to picture on cover].
Heather: Oh, snowman.
Mother: So what do you think that book's about? Hmm? Do you know? [Heather opens book]. What's that?
Heather: Big egg man. [Heather traces the picture with her finger].
Mother: No, that's one big snowball. [Mother points to the first word. Heather turns the page].
Heather: Two. [Heather points to picture].
Mother: Two what?
Heather: Two snowballs.
Mother: Two snowballs. [Heather turns page]. Ohh, what was put on there? [Mother points to picture].
Heather: Nose, eyes. [Heather points to each in the picture].
Mother: Eyes and nose. [Mother points to print as she corrects child].
Heather: Eyes and nose. [Heather turns the page].
Mother: Ohh, what else?
Heather: A smile.
Mother: A smile? [Mother checks that she heard the child correctly].
Heather: Uh, huh. [Heather turns the page and shrugs shoulders to indicate that she doesn't know what this page says].
Mother: What did you call him when you first opened the book, do you remember? . . . Hi, Frosty.
Heather: Hi Frosty. [Heather closes the book].

Notice how Heather uses what she knows, pictures, and her mother helps her to learn the meaning of the text. Heather's mother helps her to understand print/picture connections by pointing to the picture and then to the print. Heather's mother emphasizes the text meaning and context by asking Heather what the book is about and by reminding Heather of what she had called the snowman earlier. Heather's mother provides information about this print by correcting Heather's wording of one page. This manner of interacting is the essence of many home literacy experiences. The child is encouraged to use what she knows to learn a little more.

Through interchanges of this nature, parents can guide learners to move from a situation where the reading task is carried out with their help to one in which children operate independently (Brown & Reeve, 1985). Parents support letters, sounds, and concepts about the emergence of written language by holding children accountable for their share of the communication interchange, guiding them back to the question or subject under discussion and encouraging a meaningful discourse.

Because being literate results in different ways of thinking and reasoning (Egan, 1987), that very difference leads literate parents toward deliberate instruction. For example, mothers reading picture books to young children use the pictures and then letters and words as "naming games." Parents often simplify story language and then move gradually to more complex, book language. Parents also teach language structures and word meanings by encouraging children to talk about and define the book language. Thus, children learn concepts and terms for referring to language, with metalinguistic terms such as *word, sentence, letter, say, ask, assert, request* and metacognitive verbs such as *believe, desire, intend, mean, think, now, pretend, wonder, decide, realize, remember,* and *doubt* (Olson, 1984).

A long-term study of the relationship between family language interactions and academic success showed the importance of certain type of language talk to young children (Norman-Jackson, 1982). The study was a comparison of successful and not-so-successful readers in second grade who had younger brothers or sisters at home. When these younger siblings' language interactions with family members were compared, she found that their language was more mature if they were siblings of successful readers. Five years later it was found that their preschool family language experiences and activities predicted reading ability. "Parents and school age siblings in successful families, in contrast to unsuccessful families, gave more explanations, taught the ABC's, and played school; they discouraged their verbalizations less" (p. 356).

Language interactions in the context of book reading blend the new with the old—letter sounds with language sounds, text with pictures, and written language with oral language. We now show how these experiences foster literacy transfer.

HOW LITERACY IS FOSTERED THROUGH READING TO CHILDREN

Children who begin to develop literacy prior to schooling do so because parents play a direct role in teaching these beginning dispositions and abilities. Storybook reading may be the principal mechanism for literacy transfer.

Book-reading sessions help children form constructs about written language, according to Snow (1983), by highlighting "decontextualized language" features. These features include acquainting children with an impersonal author, a distant setting, varying points of view of characters, and written language forms. Studying her son's early language acquisition, Snow found the routine of book reading ideal for learning language. Because the events of each page are the same each time the book is opened, and can be used for discussion, the child can gradually take on more and more of the adult role in the reading situation. The child is provided with the chance to learn the rules for reading and acquire knowledge of written language features.

Snow and Ninio (1986) analyzed videotaped sessions of parents reading to their children and found that children learn many basic concepts about books through joint picture book reading sessions. In their talk with children about books, parents acquaint children with written language features. That is, parents acquaint children with an impersonal author, a distant setting, varying points of view of characters, and written language forms. Book reading can be an ideal setting for language learning because the events of each page are the same every time the book is opened, enabling children gradually to take over the role of telling the story, talking about it, making interpretations, and eventually reading. Children learn some of the limits of written representations, relevance of symbols, and how objects written about are represented and related to their world. They realize that, although pictures and text information are static, they can represent entire events and complex ideas. Thus, they learn that books create an autonomous fictional world in which events occur, not with real objects of the world, but with real-world social conventions and outside of real time.

A longitudinal study by Wells and associates also found that parents transmit literacy through book reading. A group of 32 children were followed from age 18 months until 10 years, with data collected on parent–child conversations, parent interviews, child interviews, and assessments by teachers and researchers. Moon and Wells (1979) revealed high correlations between reading at age 7 and preschool knowledge of literacy, parent interest in literacy, and parent quality of feedback and richness of interaction with their child. Wells (1981) showed, in addition, moderate correlations between reading at age 7 and the following: children's language competency before age 4, a range of preschool activities, preschool interest and concentration on literacy, and language skills at age 5. Wells (1986) indicated that listening to a story read aloud during parent conversations at age 3 was significantly associated with oral language ability and knowledge of literacy at age 5 and reading comprehension at age 7 and age 10. Thus, a convincing argument is made of the importance of parent–child discourse and story reading activities during the preschool years.

Studying language acquisition of 6 to 10-year-olds, Chomsky (1972) found a strong relationship between children's exposure to storybooks and their rate of linguistic development. Furthermore, kindergarten children who had been exposed to many books that were linguistically complex understood more complex language. To exemplify the relationship, Chomsky describes two kindergarteners of the same age and IQ. The child with a low linguistic comprehension score had not heard a single story during the previous week, according to parents' report. She was typically read to less than one-half hour per week and was familiar with very few complex stories. In contrast, a child with a high linguistic score had been read 17,500 words that week, was usually read to more than 2 hours a week, and frequently heard complex stories. Chomsky concludes that there is a distinct language advantage for children if they are read to frequently and hear a variety of rich and complex stories. Purcell-Gates (1988) also found that children who are read to frequently expect books to be more syntactically integrated. In addition, they can accept word orders such as inversion of the subject and a verb ("Hiding was a dragon.") that occasionally appear in books but seldom are used in oral conversations.

Close observation of parent–child book reading sessions suggests that the quality of the interaction is valuable. Ninio and Bruner (1978), who analyzed mothers' dialogues that accompanied picture book reading to young children, found that mothers direct the child's attention to particular features in a book, ask questions, provide labels, and give feedback by repeating or extending the child's remarks. Harkness and Miller (1982) found that mother–child interactions change over time. Although questions or comments to initiate book-reading interactions are continued, mothers gradually increase the length of time between each interchange by reading longer text sections. Children are able to listen to stories for longer stretches of time.

Book reading affords a flexible interaction around literacy. Joint picture–book reading was studied by DeLoache and DeMendoza (1987) with 30 pairs of mothers and 12- to 18-month-old infants. They reported significant increases over age in requests by mothers for children to say or do something, in use of elaboration feedback, particularly to connect the story information with children's experiences, and in use of questions instead of labels or comments. Over the age span children became more able to respond to mothers' questions, to use verbal (compared with nonverbal) terms, and to initiate verbal interactions.

Pellegrini, Brody, and Sigel (1985) compared parent–child book-reading interactions over age (4 and 5 years) and communication status (normal or speech disability). Analyses of 120 parents who each read two stories to their child revealed an effect for communicative status but not age. Parents of speech-handicapped children made fewer story-responding demands, paraphrased greater portions of the text, and gave more turns. The authors conclude that parents adjust their interaction styles to their child's level of communicative competence.

However, not all studies find that parents effectively transfer literacy knowledge by reading to children. The amount of time spent reading may be less effective than the quality of reading (Dunn, 1981). In an observational study of low social economic status mothers reading to their children, Guinagh and Jester (1972) found great variation in mothers' reading-aloud techniques, which affected the amount of involvement of the children. Lancy (1988) also found large differences between mother–child interactions, with mothers of early readers encouraging book discussion and emphasizing comprehension and mothers of late readers focusing on word by word reading and letter–sound analysis.

What might account for these discrepancies? We believe there is another factor which affects literacy development. The next section discusses sociocultural influences of literacy and how this group of factors must be considered.

SOCIOCULTURAL FACTORS OF LITERACY

A young child's home literacy environment is affected by the income and education of the parents, which describes social class. All too often in research, social class is used as a variable to represent the quality of a child's home environment. Social class becomes the distinguishing factor for reading achievement and literacy development. However, with minorities overrepresenting the lower strata, social class will usually be confounded with culture, and a sociocultural view of literacy is necessary.

Share, Jorm, Maclean, Matthew, and Waterman (1983) analyzed the relationships among school entry children's early reading achievement, language ability, social class, and home atmosphere. Results indicate that home atmosphere and to a lesser extent social class contribute to reading achievement and to oral language. They concluded, "Families of children who have higher early reading achievement at school entry tend to engage in more literary activities, prefer educational non-commercial television viewing, have higher occupational status and higher educational aspirations for their children" (p. 85).

Iverson and Walberg (1982) reviewed 18 cross-national correlational studies of school-aged children. Ability and achievement were closely linked to the sociopsychological environment—which they defined as academic guidance, achievement, family intellectuality, work habits, and language—and to intellectual stimulation in the home, rather than to social class (based on parent occupation and amount of education). Using metaanalysis with about 100 studies, White (1982) refined the notion of the relation between social class and academic achievement. When social class is defined as income, education, or occupation of household heads, and the child is the unit of analysis, correlations of .23 are found with academic achievement. The addition of home atmosphere factors, such as reading materials in the home, family stability, aspirations of parents for their child, and the amount of cultural activities in which the family participates, increase the average correlations to .37.

We assume that children learn about written language with respect to communities' interpretations of literacy. For example, children of working-class families in Israel are less likely to have books at home or to be read to than are children of the middle class (Feitelson, Kita, & Goldstein, 1986). Middle-class homes had an average of 54 children's books whereas working-class homes had an average of only five books. Middle-class parents read to their children up to a half hour each day, beginning at age 1. By contrast, 60 percent of working-class parents did not read at all to their children. Those who did, read infrequently, and often did not begin until their child was 5 years old. Book use in the child's socio/cultural environment certainly affected these children's literacy development.

Middle class American families typically have a print-rich environment which features signs, menus, forms, directories, newspapers, regulations, instructions, letters, calendars, bills, schedules, and books. Children are regularly read to from birth, are surrounded by books and writing materials, learn to recite the alphabet, recognize and print alphabet letters, and are encouraged to use and play with literacy materials. Lower-class groups' literacy environment is also filled with print, but often without books for children.

Home observations of 3-year-old children from working-class families in the United States have determined that joint participation in literacy events is often different from the mainstream pattern, and that there are fewer occasions of parent–child story reading (Anderson & Stokes, 1984; Teale, 1986). A case study of three American children between the ages of 2 and 4, one each from the mainstream (a middle-class white family), working-class white family, and a working-class black family, revealed story-telling differences among the families (Heath, 1983, 1986). The mainstream family engaged their child in two or three times as much story talk as did the other two families. Parent–child story telling interactions in the mainstream home involved recounting (the child describing events in answer to adults' questions), accounting (the child initiating descriptions of events experienced), event casting (a running narrative by the mother of a currently occuring event,) and the telling of imaginary stories.

Differences in literacy support within social classes and cultures are described in a study of children of working-class black families (Clark, 1984) in a comparison of families of fourth-grade children who were achieving successfully in a school with those who were not. Studying the typical structure of events during a day in the life of each child, and analyzing the social organization of those individual events, Clark found that parents of successful children provided a balanced repertoire of formal and informal learning activities and used authoritative communication to reach a consensus. They helped the child set aside time frames for the activities and provided materials and an appropriate physical climate. They also started and monitored a homework activity, deliberately instructing the child by telling or demonstrating how to do it, letting the child try it, asking questions and providing corrective feedback, praising whenever the child preforms according to the rules of the situation, and reprimanding when necessary.

Conceivably, social class and culture mark different opportunities to experience literacy activity at home. If so, it could be supposed that opportunities to experience literacy activities and talk about reading and writing with parents affect children's success in learning to read. In support of this idea, Dickinson and Snow (1987) found that the social class of the family differentiated kindergarten children's scores on printing letters and words, letter and word recognition, concepts about how and where to read, listening comprehension, and formal word definition but not on their ability to identify signs and labels in context, label pictures, or describe pictures.

Overall, then, child's developing literacy concepts that are linked to sociocultural differences appear to influence literacy acquisition. In some cases, parents' support for literacy is different from what is needed for success in schools as schooling is now structured. If the role of sociocultural factors is ignored under these circumstances, too many children who differ from the mainstream society and culture will continue to fail in school which in turn will result in lower levels of literacy in adulthood and perpetuates the cycle for the next generation. We wonder, then, whether the activity of reading to children in the home can help to break this negative cycle.

INTERVENTION RESEARCH

In a set of three studies McCormick and Mason (1983, 1986, 1989) worked with children of working class white families to determine whether making books available at home would affect children's early reading. The first study determined that, if children were taught at school how to read little books (the books were six- to eight-page picture-phrase books with simple line-drawn illustrations), and the materials were sent home afterwards, parents used the materials at home with their children. Parents also reported that they felt they were helping their child learn to read.

In the second study effects of home book reading were investigated by sending the little books home to one group of kindergarten children but not the other. The experimental group received in the mail eight books to read at home after one book had been read at school. The remaining books were to be read with parents' help at home. The matched control group received no materials. A brief letter described to parents how to read them to their child. At the end of the kindergarten year, tests of children's story-reading spelling and reading words out of context revealed superior reading performance among children who had received the books, with benefits being maintained into first grade.

In the third study, school treatments were provided to both the experimental and control subjects before the materials were sent home. The school treatments, given to children while they were in Head Start, were 10 lessons of books for the experimental group to read and 10 lessons of stories with picture materials for the control group. Experimental subjects received the books at home during the first year and an additional set during kindergarten. Control subjects received the same number of mailings at home, but the materials were pictures during the first year of the study and visual perception activities in kindergarten instead of books. Materials were given to both groups so that all children would receive something in the mail. Significant differences found at the end of the Head Start and kindergarten years, and again at the end of first grade, showed children in the experimental group with higher scores than control children on story reading, word reading, and spelling.

To understand why the materials made a difference, Kerr, Mason, and McCormick (1991) videotaped 18 parent–child dyads while reading one of two types of early literacy materials. The little books were contrasted with another type of successful early literacy materials, rhymes. Families received materials once a month for 9 months.

Analyses of videotapes indicated that mothers receiving the little books provided more appropriate basic concepts about print and oral–written language distinctions, engaging in interactions that typify those considered successful among middle-class mothers and children. They helped children develop concepts about print by pointing to picture and print as they read, and they had children read the text after they had done so, hold the book, and turn the pages. These mothers also fostered story language awareness by asking questions about the text ideas and by expanding on these ideas. They helped the children see the connections between their own experiences and the text, and rephrased as necessary to keep the text meaningful. It was apparent that low-income mothers can support their children's emerging literacy but are better able to do so when they are given appropriate literacy materials.

Edwards (1989) worked with lower-class black mothers and their kindergarten children in rural parishes of Louisiana. In her first study, she met with the mothers and demonstrated how to read to their children. Then she videotaped them as they did so, and replayed the videos to talk about how they might improve their book reading techniques. This turned out to be very threatening to several of the mothers. Low literacy levels among the mothers made it difficult for them to read to their child, and their overly strict parenting techniques often prevented them from constructing pleasurable and effective book-reading interactions with their child. In her second study, Edwards provided more community and school support for book reading. A beginning step was to work with mothers in groups, offering them opportunities and advice about reading to each other before they read to their child. With more assistance initially, parents became more comfortable with books and were more motivated to develop their own literacy skills and be more successful in helping their child enjoy books.

In both of these studies, interactions around book reading over time allowed children to focus on the print as they heard and talked about a story, to relate text events to their own experiences, and to take over some of the reading activity. When they discussed a text with their mother, they began to predict words, phrases, and story events; take the initiative in questioning; elaborate on text information; recite the text until they could repeat it without help; and negotiate with their mother about increasing their role in the reading activity. One 4-year-old Louisiana child, according to Edwards, even began a neighborhood crusade, reading books to children younger than herself.

The role of the adult in these book reading sessions, after guided practice in group or individual sessions, approximates Vygotsky's (1978) notion of supporting parents in ways that allow them to support their children. At both levels, literacy support as well as guidance are required to allow learners gradually to take on further responsibilities. While progress is slow, there appear to be not-too-costly ways to help mothers use book reading as a vehicle for literacy development. Mothers can learn (a) to model appropriate comprehension activities while reading a book to their children, (b) to make some book-reading strategies visible to the children, (c) to engage in an "online" diagnosis of difficult portions of the text and notice whether their children understand it, (d) to give appropriate feedback to comments and questions, and (e) to ask a little more of children as they acquire competence. Research in large school districts and among other cultural groups is still needed, but we believe this is an appropriate direction into literacy enrichment.

CONCLUDING REMARKS

Egan (1987) traces the path of orality to literacy, through a historical and philosophical argument, beginning with the Homeric period, which features orality, into the Platonic, in which literacy rises. He points out that an oral stance is poetic, an energetic immersion into ways to enchant, enthrall, and impress hearers. It features metaphor, rhyme, rhythm, and story telling as ways to involve and improve memory. It stresses participation in the here and now, is resistant to change, and espouses learning through experiences. A literate stance, by contrast, is a rational and critically reflective enterprise. It engages the mind in mental analysis rather than concrete concepts, focuses on linear (rather than interactive) ideas, and leads to the development of abstract concepts. Western culture is embedded in this rational view, in knowing, with feelings and emotions treated as separate mental capacities, divorced from systematic methods of learning.

Egan asserts that Western culture, having inadvertently turned away from the benefits of orality, may have bred illiteracy, an absence of both oral and literate qualities. He proposes that early childhood education attend to "prominent features of orality—the poetics of memory, participation and conservation, and classification and exlanation—that have proved *bon á penser* [good for thinking]" (p. 449). Features of orality that could be preserved and coupled with children's growth toward literacy include the following characteristics: sensitivity to the sounds of words and their emotional effects, a precise use of metaphor, rhythm, rhyme, repetition, story-shaping of narratives and myths with vivid characters and events, visual imagery, close and systematic observation of natural objects and processes, development of a sense of humor and of the absurd, and a use of binary opposites to catch and expose important themes.

We believe that book reading, story telling, memorizing songs and rhymes, and reciting the alphabet maintain strengths of the oral tradition and simultaneously promote literacy. We may not have adequately utilized or appreciated these time-honored activities. It is time to revive and expand on them.

Spencer (1986) offers a commentary about the difficulty of collaborating with parents to foster literacy at home. She observed a parent–teacher meeting in a multicultural, multilingual primary school in a socially mixed district of inner London where teachers want to convince parents to read to their children at home. They present a videotape showing family groups and children sharing books, writing, and looking at print in varying contexts. The parents agree that reading is "basic" to learning, but they are uncomfortable about the notion that reading could occur in a pleasurable context. They assume that reading must be work.

To convince parents that "children's play at its most significant is as effortful as work if the goal is desired . . . [teachers] propose that reading, like television, can be both instructive and recreational, that narratives are good continuous text, that children can wait more than a day without reading to a teacher if the teacher then gives each child a sustained hearing, that a test score is not the only measure of progress" (p. 450). The truce is an uneasy one, stemming from divergent perspectives about literacy.

Yet our studies about the process of learning to read as well as the evidence about adult–child interactions surrounding story-book reading allow the conclusion that parent–child book reading can be effective. Through reading and talk about books, parents can lead their children into important insights about how to read and comprehend written information. Book reading can build on orality while allowing children to refine and extend their concepts about written language. It can help them form a bridge from oral to written language so that they begin to analyze the distinguishing features of written language. It can give them a foundation upon which to construct meaningful representations of written texts. Despite the many pitfalls, we recommend expansion of early home and school progams that utilize the strengths of orality as well as the connections between oral and written language. Finally, parents should be assisted, as needed, so that they become avid storybook readers to their young children.

REFERENCES

Adams, M. (1990). *Beginning to read: Thinking and learning about print.* Cambridge, MA: MIT Press.

Anderson, A. B., & Stokes, S. J. (1984). Social and institutional influences on the development and practice of literacy. In H. Goelman, A. Oberg, & F. Smith (Eds.), *Awakening to literacy* (pp. 24–37). Exeter, NH: Heinemann.

Barrett, T. (1965). The relationship between measures of prereading, visual discrimination and first grade reading achievement: A review of the literature. *Reading Research Quarterly, 1,* 51–76.

Bradley, L., & Bryant, P. (1985) *Children's reading problems.* New York: Basil Blackwell.

Brown, A. L., & Reeve, R. A. (1985). *Bandwidths of competence: The role of supportive contexts in learning and development* (Tech. Rep. No. 336). Urbana: University of Illinois, Center for the Study of Reading.

Carroll, J. B. (1987). The national assessment in reading: Are we misreading the findings? *Phi Delta Kappa, 68*(6), 424–430.

Chafe, W. (1982). Integration and involvement in speaking, writing, and oral literature. In D. Tannen (Ed.), *Spoken and written language: Exploring orality and literacy, Vol. IX. Advances in discourse processes* (pp. 35–53). Norwood, NJ: Ablex Publishing Corp.

Chafe, W. (1985). Linguistic differences produced by differences between speaking and writing. In D. Olson, N. Torrance & A. Hildyard (Eds.), *Literacy, language, and learning.* Cambridge, UK: Cambridge University Press.

Chomsky, C. (1972). Stages in language development and reading exposure. *Harvard Educational Review, 42*(1), 1–33.

Clark, R. (1984). *Home interaction and children's school learning.* Unpublished manuscript, Claremont Graduate School, Claremont, CA.

Clay, M. (1979). *Reading: The patterning of complex behavior.* Portsmouth, NH: Heinemann.

DeLoache, J. S., & DeMendoza, O. A. P. (1987). Joint picturebook interactions of mothers and one-year-old children. *British Journal of Developmental Psychology, 5,* 111–123.

Dickinson, D., & Snow, C. (1987). Interrelationships among prereading oral language skills in kindergarten from two social classes. *Early Childhood Research Quarterly, 2*(1), 1–25.

Downing, J., Ayers, D., & Schaefer, B. (1978). Conceptual and perceptual factors in learning to read. *Educational Research, 21,* 11–17.

Dunn, N. (1981). Children's achievement at school-entry as a function of mothers' fathers' teaching sets. *The Elementary School Journal, 81*, 245–453.

Durrell, D., & Murphy, H. (1953). The auditory discrimination factor in reading readiness and reading disability. *Education, 73*, 556–560.

Dyson, A. H. (1984). Emerging alphabetic literacy in school contexts: Toward defining the gap between school curriculum and child mind. *Written Communication, 1*(1) 5–55.

Edwards, P. (1989). Supporting lower SES mothers' attempts to provide scaffolding for book reading. In J. Allen & J. M. Mason (Eds.), *Risk makers, risk takers, risk breakers* (pp. 222–250). Portsmouth, NH: Heinemann.

Egan, K. (1987). Literacy and oral foundations of education. *Harvard Educational Review, 57*, 445–472.

Ehri, L. (1983). A critique of five studies related to letter-name knowledge and learning to read. In L. Gentile, M. Kamil, & J. Blanchard (Eds.), *Reading research revisited* (pp. 143–154). Columbus, OH: Merrill.

Ehri, L., & Wilce, L. (1987). Does learning to spell help beginners learn to read words? *Reading Research Quarterly, 22*, 47–65.

Feitelson, D., & Goldstein, Z. (1986). Patterns of book ownership and reading to young children in Israeli school-oriented and non school-oriented families. *The Reading Teacher, 39* (9), 924-930.

Gates, A., Bond, G., & Russell, D. (1939). *Methods of determining reading readiness.* New York: Columbia Teachers College.

Goswami, U., & Bryant, P. E. (1987). Rhyming, analogy, and children's reading. In P. B. Gough (Ed.), *Reading acquisition.* Hillsdale, NJ: Erlbaum.

Guinagh, B., & Jester, R. (1972). How parents read to children. *Theory into Practice, 11*, 171–177.

Harkness, F., & Miller, L. (1982). *A description of the interaction among mother, child, and books in a bedtime reading situation.* Paper presented at the seventh annual Boston Conference on Language Development, Boston, MA.

Heath, S. B. (1983). *Ways with words: Language, life and work in communities and classrooms.* New York: Cambridge University Press.

Heath, S. B. (1986). Separating "things of the imagination" from life: Learning to read and write. In W. Teale & E. Sulzby (Eds.), *Emergent literacy: Writing and reading* (pp. 156–172). Norwood, NJ: Ablex Publishing Corp.

Iverson, B., & Walberg, H. (1982). Home environment and school learning: A quantitative synthesis. *Journal of Experimental Education, 50*, 144–151.

Kerr, B. M., Mason, J. M., & McCormick, C. (1991). *Literacy discourse in shared reading events* unpublished manuscript. Urbana: University of Illinois, Center for the Study of Reading.

Lancy, D. F. (1988). *Parents' strategies for reading to and with their children.* Paper presented at the AERA annual meeting.

Lartz, M., & Mason, J. (1988). Jamie: One child's journey from oral to written language. *Early Childhood Research Quarterly, 3*, 193–208.

Lomax, R. G., & McGee, L. M. (1987). Young children's concepts about print and reading: Toward a model of word reading acquisition. *Reading Research Quarterly, 22*(2), 237–256.

Lundberg, I. (1986, March). *Two dimensions of decontextualization in reading acquisition.* Paper presented at the Conference on Early Reading, Center for Cognitive Science, University of Texas at Austin.

Mason, J. (1980). When do children begin to read? An exploration of four year old children's letter and word reading competencies. *Reading Research Quarterly, 15,* 203–227.

Mason, J. (1984). Early reading from a developmental perspective. In P. D. Pearson (Ed.), *Handbook of reading research.* New York: Longman.

Masson, M. (1985). Rapid reading processes and skills. In G. E. MacKinnon & T. G. Waller (Eds.), *Reading research, advances in theory and practice* (Vol. 4). New York: Academic Press.

McCormick, C., & Mason, J. (1983). *Intervention procedures for increasing preschool children's interest in and knowledge about reading* (Tech. Rep. No. 312). Urbana: University of Illinois, Center for the Study of Reading.

McCormick, C., & Mason, J. (1986). Intervention procedures for increasing preschool children's interest in and knowledge about reading. In W. Teale & E. Sulzby (Eds.), *Emergent literacy: Writing and reading* (pp. 90–115). Norwood, NJ: Ablex Publishing Corp.

McCormick, C., & Mason, J. (1989). Use of little books at home: A minimal intervention strategy for fostering early reading. In J. Allen & J. Mason (Eds.), *Risk makers, risk takers, risk breakers.* Portsmouth, NH: Heinemann.

Monroe, M. (1932). *Children who cannot read.* Chicago: University of Chicago Press.

Moon, C., & Wells, G. (1979). The influence of home on learning to read. *Journal of Research in Reading, 2,* 53–62.

Ninio, A., & Bruner, J. S. (1978). The achievement and antecedents of labeling. *Journal of Child Language, 5,* 5–15.

Norman-Jackson, J. (1982). Family interactions, language development, and primary reading achievement of Black children in families of low income. *Child Development, 53,* 349–358.

Olson, D. (1984). "See! Jumping!" Some oral language antecedents of literacy. In H. Goelman, A. Oberg, & F. Smith (Eds.), *Awakening to literacy.* Portsmouth, NH: Heinemann.

Pelligrini, A., Brody, G., & Sigel, I. (1985). Parents' book-reading habits with their children. *Journal of Educational Psychology, 77*(3), 332–340.

Perena, K. (1984). *Children's writing and reading: Analyzing classroom language.* Oxford, UK: Basil Blackwell.

Perfetti, C. (1986). *The representation problem in reading acquisition.* Paper presentation at the Conference on Early Reading, Center for Cognitive Science, University of Texas.

Purcell-Gates, V. (1988). Lexical and syntactic knowledge of written narrative held by well-read-to-kindergartners and second graders. *Research in the Teaching of English, 22*(2), 128–160.

Share, D., Jorm, A., Maclean, R., Matthew, R., & Waterman, B. (1983). Early reading achievement, oral language ability, and a child's home background. *Australian Psychologist, 18,* 75–87.

Snow, C. (1983). Literacy and language: Relationships during the preschool years. *Harvard Educational Review, 53,* 165–189.

Snow, C., & Ninio, A. (1986). The contribution of reading books with children to their linguistic and cognitive development. In W. Teale & E. Sulzby (Eds.), *Emergent literacy: Writing and reading* (pp. 116–138). Norwood, NJ: Ablex Publishing Corp.

Spencer, M. (1986). Emergent literacies: A site for analysis. *Language Arts, 63,* 442–453.

Stanovich, K. (1986). Matthew effects in reading: Some consequences of individual differences in the acquisition of literacy. *Reading Research Quarterly, 21,* 360–407.

Stanovich, K., Cunningham, A., & Cramer, B. (1984). Assessing phonological awareness in kindergarten children: Issues of task comparability. *Journal of Experimental Child Psychology, 38,* 175–190.

Stedman, L., & Kaestle, C. (1987). Literacy and reading performance in the United States, from 1880 to the present. *Reading Research Quarterly, 22,* 8–46.

Tannen, D. (1982). Oral and literate strategies in spoken and written narratives. *Language, 58,* 1–21.

Teale, W. H. (1986). Home background and young children's literacy development. In W. H. Teale & E. Sulzby (Eds.), *Emergent literacy: Writing and reading* (pp. 173–206). Norwood, NJ: Ablex Publishing Corp.

Torrance, N., & Olson, D. R. (1985). Oral and literate competencies in the early school years. In D. R. Olson, N. Torrance, & A. Hildyard (Eds.), *Literacy, language and learning* (pp. 256–284). Cambridge, MA: Cambridge University Press.

Treiman, R. (1986, March). *The role of intrasyllabic units in learning to read and spell.* Paper presented at the Conference on Early Reading, Center for Cognitive Science, University of Texas at Austin.

Tunmer, W. E., Herriman, M. L., & Nesdale, A. R. (1988). Metalinguistic abilities and beginning reading. *Reading Research Quarterly, 23,* 134–158.

Vygotsky, L. S. (1978). *Mind in society: The development of higher psychological processes.* Cambridge, MA: Harvard University Press.

Wells, G. (1981). Some antecedents of early educational attainment. *British Journal of Sociology of Education, 2,* 181–200.

Wells, G. (1986). *The meaning makers.* Portsmouth, NH: Heinemann.

White, K. (1982). The relation between socioeconomic status and academic achievement. *Psychological Bulletin, 91,* 461–581.

Yopp, H. (1988). The validity and reliability of phonemic awareness tests. *Reading Research Quarterly, 23,* 159–177.

Chapter 5

Forward to the Past: Black and White American Families, Literacy, and Policy Lag*

Diana T. Slaughter–Defoe
Northwestern University

This invited critique addresses a number of both explicit and implicit theoretical assumptions in the paper by Mason and Kerr. Four of the major assumptions are:

1. The home environment significantly influences early learning and cognitive development;
2. Parental involvement in programs addressing children's early learning and development can significantly help lower-income white and black children acquire literacy and therefore experience greater success with traditional schooling;
3. Enough is known about the course of literacy development in both children and adults such that appropriate instructional goals and strategies can be designed to remedy individual developmental deficits;
4. Enough is known about who should be targeted by adult–child literacy programs, what factors are likely to have precipitated their deficits, and how to deliver these literacy programs to persons of diverse sociocultural backgrounds through networking and negotiations within the American political process.

---

* A version of this paper was given at the annual meeting of the American Educational Research Association, New Orleans, LA, April 5–9, 1988.

I argue that, in 1988, only the first assumption is valid. The second assumption deserves qualified positive support, but the third and fourth assumptions are patently false. For contemporary research on early literacy to be useful to the lower income families who need it most, including racial and ethnic minority families, such as blacks, investigators in this area must build better bridges to the social and educational policy arena.

Lessons can be learned from the field of child development research; we know considerably more about the growth and development of children than we are able to implement, primarily because ties between the scientific child development community and the social and public policy arena are so tenuous. Our major child policy program, Project Head Start, for example, has never served over 25 percent of the eligible national disadvantaged population in any given year, and we have only indirect evidence of its sphere of influence as a demonstration program. Further, and ironically, regarding the potential impact of the findings in reference to early literacy, adult literacy is a particular barrier to reaching children in impoverished communities and neighborhoods. Parents in these communities are least likely to turn to the printed media for information about childrearing and child care, or childhood growth and development. Therefore, it is incumbent upon the investigators themselves to devise strategies for ensuring that the benefits of their research reach those segments of the American public that are simultaneously most isolated from mainstream information and, probably, most needful of that information. Research and policy efforts in the mid- to late-1960s on parenting influences upon cognitive development and academic achievement will be contrasted with contemporary efforts to relate family influences to the development of literacy skills.

CHILD DEVELOPMENT RESEARCH SINCE THE 1960s: PARENTAL INFLUENCES ON EARLY LEARNING AND COGNITIVE DEVELOPMENT

Since the renewed interest in early childhood education and child development in the early 1960s, developmental researchers have clarified much about the development of young children and about what constitutes quality child care in primary group settings, including families, nurseries, day care settings, preschools, and kindergartens. Typically, earlier theoretical positions have been affirmed, but new insights have been generated by research. Like Freud, researchers have affirmed the importance of primary family relationships to the children's development. Higher-order mammalian infants are not born with preadapted response patterns that appear at predetermined times and persist throughout life. Rather, primary social relationships are so important that infants seem preadapted to promote bonding between themselves and their earliest caregivers such that they will be protected while they learn during the many months, even years, of dependency. Further, it now seems universal that normal human infants seduce their adult caregivers into what can become a very mutually satisfying relationship. From the child's perspective, the relationship is initially dependent upon the close proximity of the caregiver. Later, around 18 months, when children develop a sense of object permanence and stability, that bond or tie is no longer dependent upon the presence of the caregiver, but can endure through enormous distances and times.

Like Piaget, American developmental researchers have affirmed the active roles that children take in the construction of their knowledge of the world. Thanks to research with infants, we know that even newborns can learn. For example, they can be conditioned to respond differently to sounds and voices. However, with respect to cognitive development, we have also found that context importantly influences the rate of children's development, what children can learn, and how they perform on conceptual tasks. Findings support some key Piagetian assumptions, but in many instances, remarkable new information has evolved.

In summary, when one of our 3-year-old preschool children seizes a toy telephone, attempts to dial, and insists that she will "call Mommy," we understand the developmental import of that complex behavioral pattern. We understand that the behavior reflects the convergence and integration of complex cognitive and social skills in the service of an immediate emotional need. We also understand that the acquisition of such skills was not solely spontaneous, random, or even imposed by well-intentioned others. Rather, it was gradually, systematically, and quite deliberately won from a negotiated relationship between the child and her earliest caregivers.

With regard to early literacy, including reading and writing, it seems that the components assume predictable developmental sequences that can be described and charted well before children reach first grade. The same is true of number concepts. Many good teachers have intuitively identified these patterns and then accommodated their teaching goals and methods. In the future, there is the promise that all teachers can learn to analyze student responses and adapt their instructional strategies to each student's developmental status in a given curriculum area. Models for studying these complex issues were pioneered in the 1960s and 1970s. Even more remarkable, we have broadened our definition of context as it applies to children's learning. We now appreciate that, beyond nurseries, preschools, and kindergartens, children can and do learn in day care centers, nonfamily day care homes, and the homes of extended family members such as grandparents, as well as their own homes. They learn from parents, grandparents, older siblings, and other relatives, surrogate and ancillary child care providers, and even peers!

A definite relationship between children's early home environment and language development has been found. For example, parental teaching styles have been shown to be related to more successful performance on standard intelligence and achievement tests, particularly between ages 12 to 60 months. Parental influences on children's cognitive development have been observed as early as 3 months. When mothers respond to babies' specific vocalizations, and when a process of "turntaking" ensues as if a conversation were occurring, children's language development is positively affected. By age 1, the primary caregiver's crucial role is even more apparent. Nelson (1973), Clarke-Stewart (1973), and Carew (1980) have each shown in their research that, between 12 and 36 months, the child's advances in, for example, language development are stimulated by maternal verbal behaviors that are contingently responsive to the child's speech behaviors, and by verbal and nonverbal behaviors indicative of active participation in the child's ongoing play. Similar findings have been obtained in a series of research studies conducted with young black children by Caldwell and her colleagues (e.g., Elardo, Bradley, & Caldwell, 1975; Bradley & Caldwell, 1977, as well as, most recently, by Blumenthal, 1987). Schachter (1979) has argued that the value of verbal responsivity, from the child's perspective, is not primarily intellectual, but social and emotional. The mother's behaviors increase the child's sense of security and self-worth. Therefore, from such a predictable, intimate, ongoing relationship the child develops confident expectancies of impacting and controlling its human environment.

Applied developmental research, such as early intervention research (e.g., Andrews, Blumenthal, Bache, & Wiener, 1982; Slaughter, 1983) has also demonstrated that parental teaching behaviors are important to the early learning and development of children. In my research, for example, I found that, when the black mothers in my study networked to form a new "extended family" support group to one another, the children benefitted through increased maternal investment in their ongoing play behaviors. The individuation of the young children was also paralleled by a maintenance of average IQ levels during the age period (18–44 months) when levels for black children from lower-income communities typically decline to below average.

In summary, both basic and applied developmental research have demonstrated that the early home environment has a crucial impact upon the cognitive development of young children (See, for example, Gottfried, 1984; Laosa & Sigel, 1982; Rogoff & Wertsch, 1984; Scott-Jones, 1984; Snow & Ferguson, 1977; Tizard & Hughes, 1984), including young black children (Slaughter & Epps, 1987). I would also argue that the research has clearly suggested how the early learning and development of these children can be supported—at least to the extent that any research involving human beings can produce findings which are both reliable and credible. For the preceding reasons, I believe that Mason and Kerr's summary of the findings demonstrating ties between the home environment, particularly parenting and early achievement, especially early literacy acquisition, to be both compelling and accurate.

In the preceding discussion, I alluded to how early intervention programs for parents and children have been shown to be useful supports to parenting. The child advocates who often design and implement such programs understand that, in order for children to develop optimally, parents may have to change themselves and what they would "naturally" do with their children. However, they also know that, if social supports are not provided to the parents for their efforts at change, little enduring impact can be expected, even given the best of programs. Project Head Start is a federally sponsored preschool program in which parent involvement, and support for parent involvement, has been a key component (Slaughter, Washington, Oyemade, & Lindsey, 1988). What lessons have been learned from this nearly 25-year-old program?

HEAD START AND PARENT INVOLVEMENT

From the beginning, Project Head Start had a mandate to significantly involve parents in its educational programming for young children (Zigler & Valentine, 1979). This policy was initially buttressed by the child development studies of Hess and Shipman (1965) and quickly followed by a host of others, some of which have been cited earlier in this chapter. Importantly, the mandate for parental involvement in Head Start appears to have stemmed from community-action-oriented guidelines for involving the poor in decisions affecting the lives of their families (Valentine & Stark, 1979; Slaughter & Kuehne, 1988). A broader mandate for political participation, over time, became one in which parents were offered opportunities to participate in a variety of education programs designed to help parents learn to assist young children's early cognitive development. Meanwhile, the parent involvement component of Head Start has just recently begun to receive systematic attention (Washington & Oyemade, 1987), though it has been widely recognized for some time that, when parents are involved in early intervention programs, children are more likely to profit from them (Bronfenbrenner, 1976; Honig, 1980). Thus, research and policy in relation to parent involvement and children's early learning developed in parallel fashion, primarily because of the zeitgeist of the times—times in which American society felt itself more generous towards its poor and disenfranchised—and not because child development researchers and social and educational policy makers, for the most part, perceived themselves to be working together, with shared goals and objectives. I would argue that the best case of proof for my position is that, only later, after the early research evaluations of the program were less than promising, did the developers of Project Head Start acknowledge the need for mechanisms to ensure program continuity once children entered the primary grades within traditional public schools (Washington & Oyemade, 1987; Zigler & Valentine, 1979).

In fact, at least one author (Washington, 1988) has recently argued that the field of child development and social policy emerged from these parallel processes, for example, research into parental influences upon child achievement behaviors and structured efforts to include parents in ongoing daily operations of child care programs. Elsewhere, I have argued (Slaughter, 1982) that, in promoting child development research and more advanced instructional efforts in early childhood education, Project Head Start accomplished everything except to address its principal mandate: To serve as a solution to the problems of poverty in America. This is a related important issue not directly addressed by Mason and Kerr; just what, in 1988, are we trying to do? If, for example, our efforts are designed to encourage literacy acquisition in America, that is one goal. But if we really want to improve the lives of the poor and disenfranchised by promoting literacy acquisition such that these children and their parents can significantly improve the overall social quality of their lives, then that is a considerably more complex issue that merits open discussion and analysis, and particularly in view of the state of knowledge in the field of literacy development itself.

RESEARCH IN THE EIGHTIES: FAMILIES AND EARLY LITERACY DEVELOPMENT

Stubbs (1980) argues that, despite over 70 years of research, no predominant theory of reading or of instruction on how to teach reading has emerged. Further, definitions of what constitutes literacy also vary. Authors agree that literacy is concerned with both spoken and written language, and that any theory of literacy must discuss how these two phenomena are integrated in the cognitions of socially and maturationally diverse individuals. There is also agreement that the effective, efficient use of print is the key to reading achievement, though opinions vary about whether phonics and other technical aspects of using print are more crucial than comprehension of meaning, particularly in decontextualized situations (Snow, 1983).

Despite the ambiguities associated with definitions of *reading* and of *literacy,* to say nothing of the problems of measurement of reading ability (Brown, Palincsar, & Purcell, 1985), researchers continue to observe the relationship between measures of reading and of literacy and other social and psychological variables. In the 1980s, the positive and significant correlation between measures of familial social status and of academic achievement in modern industrial societies (e g., Hess, 1970; Stubbs, 1980; White, 1982) continues to invite investigation and speculation about the relationship between cultural-familial lifestyles and children's cognitive development. This time, however, the focus is more specifically on children's literacy development, a strong interest shared by anthropologists, linguists, and educators with special interests in language arts.

The researches of Brice-Heath (1982, 1983), Taylor (1983, 1986), Teale (1986), and others each point to the significance of the home and family environment to children's literacy. Most of this research attempts to describe how cultural factors (e.g., social structural and ethnicity) are mediated by family processes. The researchers assume that all children experience literacy in their home environments, but that the task of the researcher is to describe the many varieties of literacy experiences children may have. Most of these experiences are assumed to be highly associated with social participation in family and community, and relatively fleeting, casual events. However, some events, such as storybook reading in middle-class homes, are more enduring and regular, and, therefore, somewhat more amenable to systematic study. In fact, however, the more naturalistic, graphic, humanistic, and highly idiosyncratic the research methods used (e.g., Taylor, 1986), the more compelling the finding of ties between the home environment and children's acquisition of literacy.

Anderson and Stokes (1984) summarize the state of the art in this area in reference to ethnic minority families quite well:

> from the beginning we thought we would find that culture exerts a significant influence on a child's development of literacy . . . even within our lower-class sample. . . . However, when comparing the patterns of

> literacy practice presented by the three ethnic groups in our sample, we found it difficult to conclude that ethnicity was a uniformly significant source of differences. . . . It became increasingly clear over the course of the study that many of the businesses and institutions of society exert a strong influence on literacy practices of low-income people. Aside from using print to carry out official and routine activities, social institutions also were involved in the recreational activities of the people who participated in the study. . . . In the United States, the production of print for entertainment purposes can be a very profitable enterprise. . . . When literacy is equated only with books, the research reports indicate that lower-class families engage much less frequently in literacy activities than do middle-class families. (pp. 34–36)

Thus, Anderson and Stokes argue that the disposition to certain types of literacy events, namely, family interactions around schoolbooks and homework, may be more tied to the constraints of socioeconomic status than to either cultural or ethnic variations in the amounts and types of exposure to literacy events. This disposition is what seems most relevant to children's literacy development, insofar as this development enhances, and is enhanced by, academic studies. As persons within the literacy field become increasingly intrigued with the complex issues associated with studying the practical and symbolic aspects of literacy in family life (e.g., Taylor, 1983; Leichter, 1984), the viewpoint offered as a result of the authors' research increases in importance.

It is possible to look at the strengths and weakness of the Mason and Kerr paper against some of the preceding criteria. The paper presents a variety of logically distinct literacy skills each of which may be presumed to have their own course of development, but about which scientists currently appear to know very little. Further, it appears that even less is known of what the course of development is under diverse socioeconomic differences. Studies cited by Mason and Kerr tend not to use comparison groups, so we have little idea of the generalizability of findings across different social and cultural groups. Finally, there is the presumption that literacy and reading have the same relationship to one another in the different groups, when the naturalistic studies that have been conducted seem to suggest just the opposite. On the positive side, Mason and Kerr, as well as other researchers in this field, are to be commended for emphasizing the importance of naturalistic observation as a method of obtaining important, valid data. I would suggest that they continue to obtain developmental profiles of children which could be used in designing useful curricula.

Originally, however, behavioral scientists embarked on audiotape and observational methods in the study of parental influences on children's cognitive development, because we were attempting to describe parenting processes, and it seemed more sensible to collect process-oriented data to support our general conceptual orientation. Further, we thought that our obtained data would have more credibility in the policy arena. Our methods, for example, would not be quite as subject to criticisms associated with social desirability issues, nor would we be quite so removed from the ongoing everyday lives of our research subjects. Specifically, the research methods were instruments or tools to an end; the methods themselves did not dominate our considerations, and whenever they did, it was generally toward the aim of improving upon our contribution to the public policy arena.

Conversely, today scientific methodology may be reified without consideration of the ends towards which it should serve, namely, the human welfare and social justice of persons. In an otherwise excellent paper, for example, Leichter (1984) outlines six criteria for the intensive examination of families in their natural environments in their roles as educators, most of which emphasize the exploratory nature of the research, and none of which stress the potentially stronger ties of obtained data to social and educational policy issues, essentially the rationale for which this type of research began in the mid to late sixties (e.g., White & Watts, 1973).

BLACK FAMILIES, LITERACY, AND POLICY LAG

At present, it is customary to identify earlier research focusing on black families and children's literacy development with deprivation theory (e.g., see Teale, 1986). Stubbs (1980) devotes a concluding chapter to the contribution of various characteristic phases of deprivation theory to our understanding of reading failure. He argues that, whether true or not, there is little evidence in support of deprivation theory. He also argues that the theory, even if it is only a myth, can be rendered true if enough teachers and other educators believe in the "myth of the deprived child," primarily because the myth can assume the character of a self-fulfilling prophecy. I certainly agree with this position and would remind Mason and Kerr that cognitively oriented intervention programs must now begin with a different set of assumptions about poor children's intellectual capabilities than programs did nearly a quarter of a century ago. We now assume sensorimotor intelligence to be a valid form of intelligence, and that children of lower social-status families are no more likely, on average, than children of higher status families to be cognitively deficient at birth and during the first year and a half of life (Ginsberg, 1972; Slaughter, 1983; Miller-Jones, 1988). The role of early intervention with lower social status children is to enable them to continue to develop optimally by providing appropriate social and instructional supports to both the children and their responsible adult caregivers.

However, what Stubbs does not discuss, and what is only rarely discussed in more contemporary writing on this topic, is a corresponding commitment to human welfare and social justice. Many persons who essentially espoused deprivation theory (e.g., Bereiter & Engelmann, 1966) also espoused the use of social science theory and research to improve the lives of people. Consequently, and almost ironically, the proliferation of social programs in the 1960s to maximize the likelihood of equal educational opportunity on the part of black and other ethnic minorities, as well as rural, poor peoples, was closely linked to a genuine concern for the ultimate well-being of participants, particularly the children. Deprivation theorists were undoubtedly overly optimistic about the linkages between individual academic achievement and an entire community's ability to escape poverty in the latter half of the 21st century, but they were not misguided in their courageous resolve to link social science theory and research to the policy arena.

Today, the policy arena seems virtually uninformed by prior theory and research. It is instructive to digress a little to consider the situation of lower-income black families. The media project images of black children and families (cf. the 1986 Steeter documentary presented by Bill Moyers) that do not capitalize on the strengths of these children and families. Instead, the images focus on social disorganization and pathology as if the intervening years and debates in academic circles since the publication of the Moynihan report in 1965, had never occurred. Further, the media exascerbate differences between the haves and have nots in black life. The Cosby show, for example, inherently logical and consistent, only rarely depicts this black family as caring and nurturing of other black persons who are less fortunate than themselves. Despite persistent research findings indicating that ties to extended kin and communities are the defining aspects of black life in America (e.g., McAdoo, 1988; Stack, 1974; Wilson, 1986; Slaughter & Dilworth-Anderson, 1988), media today do not represent black family life this way. Rather, blacks are told that deprivation theorists have been apologists, and that they themselves must now assume major personal responsibility for their enduring poverty and blocked educational and occupational opportunities. Further, they are also being told that, since excellent parenting can occur within lower income, as well as middle income, communities, theoretical formulations about their experiences need not include consideration of the reality of social structural factors, but only of their personal perceptions and utilization of what structural resources are available. Often, the concept of social support as an important buttress to parenting is not included in the research paradigms, and, instead, the research models intensify and reify study of the particular, discrete parenting instance. There is little concern for the reliability or generalizability of obtained information, that is, little concern for the long-term social policy significance of obtained findings. Similarly, at this writing we know next to nothing of the goals and aspirations of lower income white parents; in particular, we do not know why disproportunate numbers of these families do not participate in available social programs (e.g., Head Start) for themselves and their children. Do we have any reason to believe that this pattern of behavior would change with the creation of adult–child literacy programs? Mason and Kerr's own experiences in attempting to implement their research in such a school district would suggest not, unless we pay serious attention to the processes underlying networking and negotiations with communities targeted for services (J. Mason, personal communication, April 27, 1988).

Some researchers would undoubtedly argue that it is premature to anticipate that literacy research would have a profound impact on social and educational policy. Others, applauding the caution because all the "data are not in," will also point out that policy will be most directly impacted in schools and colleges of education, because the newer findings will affect how teachers are trained. I disagree. If we can learn anything from the experience of the 1960s, it is that our approach and perceptions of social and educational policy will impact the research questions we choose, the models and paradigms within which we work, and the interpretation of obtained data. Specifically, any research which is concerned with people who endure persistent poverty, and related suffering, is inevitably enmeshed in policy considerations. Further, as I have elsewhere argued (Slaughter & McWorter, 1985), serious empirical study of black children and families arose initially in a policy-related, rather than conceptual and theoretical, context. Finally, limiting the benefits of literacy research to the education and training of adult teachers and parents constrains and restricts focus on the reality of social structural factors which play an important role in blocked educational, economic, and occupational opportunities. In a democratic state such as our own, we cannot now or ever afford to neglect these considerations.

CONCLUSION

First, in this critique I briefly discussed just how much scientific research information is available about the requirements for adaptive learning and development in early childhood. Second, I profiled, using Head Start as an example, how adequate our best efforts have been at interfacing the child development and policy arena.[1] Third, I interpreted the significance of some of the more recent findings in the area of families and early literacy, and finally, I argued that the importance of these findings is only minimally being integrated into the social and educational policy arena.

It is the irony of this situation, as just profiled, which has led me to conclude that we are rapidly moving forward into the past. Our lag in the arena of policy implementation has resulted in new and parallel subfields working on, and once again solving, a problem that, for all practical purposes, was first addressed nearly 20 to 25 years ago, specifically, the problem of how to prepare low-income black and white children for successful acquisition of school-related literacy skills. Contemporary attention to this problem, however, is not coupled with a parallel commitment to human welfare and social justice. The hard-won, virtually conventional wisdoms of the past about the important interface between research in human development and social and educational policy are only infrequently addressed in the more recent research efforts.

REFERENCES

Anderson, A. B., & Stokes, S. J. (1984). Social and institutional influence on the development and practice of literacy. In H. Goelman, A. Oberg, & F. Smith (Eds.), *Awakening to literacy*. Portsmouth, NH: Heinemann Educational.

Andrews, S. R., Blumenthal, J. B., Bache, W., & Wiener, G. (1982). The skills of mothering: A study of parent child development centers. *Monographs of the Society for Research in Child Development, 47* (6, Serial No. 198).

Bereiter, C., & Engelmann, S. (1966). *Teaching disadvantaged children in the preschool*. Englewood Cliffs, NJ: Prentice Hall.

Blumenthal, J. B. (1987). *Mother-child interaction and child cognitive competence in low income Black families: A longitudinal study*. Unpublished doctoral dissertation, The University of Chicago.

Bradley, R.; & Caldwell, B. (1977). Home environment, social status, and mental test performance. *Journal of Educational Psychology, 69,* 697–701.

Brice-Heath, S. (1982). Questioning at home and at school: A comparative study. In G. Spindler (Ed.), *Doing the ethnography of schooling: Educa-*

[1] As this chapter goes to press in 1991, the recent reauthorization of Project Head Start has included budgetary appropriations for expanded services to greater percentages of eligible children, thanks to the dedicated efforts of the current ACYF (Administration of Children, Youth, and Families, Department of Health and Human Services) Commissioner, Wade Horn, and his staff.

tional anthropology in action (pp. 96–101). New York: Holt, Rinehart & Winston.

Brice-Heath, S. (1983). *Ways with words: Language, life and work communities and classrooms.* Cambridge, UK: Cambridge University Press.

Bronfenbrenner, U. (1976). Is early intervention effective? Facts and principles of early intervention: A summary. In A. M. Clarke & A. B. D. Clarke (Eds.), *Early experience: Myth and evidence* (pp. 247–258). New York: Free Press.

Brown, A., Palincsar, A., & Purcell, L. (1985). Poor readers: Teach, don't label. In U. Neisser (Ed.), *The academic performance of minority children: A new perspective* (pp. 105–143). Hillsdale, NJ: Erlbaum.

Carew, J. (1980). Experience and the development of intelligence in young children at home and in day care. *Monographs of the Society for Research in Child Development, 45* (6–7, Serial No. 187).

Clarke-Stewart, A. (1973). Interactions between mothers and their young children: Characteristics and consequences. *Monographs of the Society for Research in Child Development, 38* (6–7, Serial No. 153).

Elardo, R., Bradley, R., & Caldwell, B. (1975). The relation of infants' home environments to mental test performance from six to thirty-six months: A longitudinal analysis. *Child Development, 46,* 71–76.

Ginsburg, H. (1972). *The myth of the deprived child.* Englewood Cliffs, NJ: Prentice-Hall.

Gottfried, A. W. (Ed.). (1984). *Home environment and early cognitive development: Longitudinal research.* New York: Academic Press.

Hess, R. (1970). Social class and ethnic influences on socialization. In P. Mussen (Ed.), *Carmichael's manual of children* (Vol. 2, pp. 457–558). New York: John Wiley.

Hess, R., & Shipman, V. (1965). Early experience and the socialization of cognitive modes in children. *Child Development, 36,* 369–383.

Honig, A. (1980). Working with parents of preschool children. In R. Abidin (Ed.), *Parent education and intervention handbook* (pp. 385–431). Springfield, IL: Charles C. Thomas.

Laosa, L., & Sigel, I. (Eds.). (1982). *Families as learning environments for children.* New York: Plenum Press.

Leichter, H. (1984). Families as environments for literacy. In H. Goelman, A. Oberg, & F. Smith (Eds.), *Awakening to literacy* (pp. 38–50). Portsmouth, NH: Heinemann Educational.

Mason, J., & Kerr, B. (1988, April). *Literacy transfer in the preschool years.* Paper presented at the Conference on the Intergenerational Transfer of Cognitive Skills, San Diego.

McAdoo, H. (Ed.). (1988). *Black families.* Beverly Hills, CA: Sage

Miller-Jones, D. (1988). The study of African-American children's development: Contributions to reformulating developmental paradigms. In D. Slaughter (Ed.), *Black children and poverty: Developmental perspectives.* San Francisco: Jossey-Bass.

Nelson, K. (1973). Structure and strategy in learning to talk. *Monographs of the Society for Research in Child Development, 38* (1–2, Serial No. 149).

Rogoff, B., & Wertsch, J. (Eds.). (1984). *Children's learning in the zone of proximal development.* San Francisco: Jossey Bass.

Schachter, F. (1979). *Everyday mother talk to toddlers.* New York: Academic Press.

Scott-Jones, D. (1984). Family influences on cognitive development and school achievement. In E. Gordon (Ed.), *Review of research in education* (pp. 259–304). Itasca, IL: Peacock.

Slaughter, D. (1982). What is the future of Head Start? *Young Children, 37,* 3–9.

Slaughter, D. (1983). Early intervention and its effects on maternal and child development. *Monographs of the Society for Research in Child Development, 48* (4, Serial No. 202).

Slaughter, D., & Epps, E. (1987). The home environment and academic achievement of Black American children and youth: An overview. *Journal of Negro Education, 56*(1), 3–20.

Slaughter, D., & Dilworth-Anderson, P. (1988, July). Care of Black children with sickle cell disease: Fathers, maternal support, and esteem. *Family Relations, 37,* 281–287.

Slaughter, D., & McWorter, G. (1985). Social origins and early features of the scientific study of Black American families and children. In M. B. Spencer, G. K. Brookins, & W. R. Allen (Eds.), *Beginnings: The social and affective development of Black children* (pp. 5–18). Hillsdale, NJ: Erlbaum.

Slaughter, D., Washington, V., Oyemade, U., & Lindsey, R. (1988, Summer). Head Start: A backward and forward look. *Social Policy Report, 3* [Available from Society for Research in Child Development, Attn.: L. Skidmore, 300 N. Ingalls, Ann Arbor, MI 48109-0406.

Slaughter, D., & Kuehne, V. (1988). Improving Black education: Perspectives on parent involvement. *Urban League Review, 11* (1–2), 59–75.

Snow, C. (1983). Literacy and language: Relationships during the preschool years. *Harvard Educational Review, 53* (2), 165–189.

Snow, C., & Ferguson, C. (1977). *Talking to children: Language input and acquisition.* Cambridge, UK: Cambridge University Press.

Stack, C. (1974). *All our kin.* New York: Harper & Row.

Steeter, R. C. (Producer & Director). (1986). *The Vanishing Family: Crisis in Black America/CBS reports* [Videorecording]. New York: Carousel Films.

Stubbs, M. (1980). *Language and literacy: The sociolinguistics of reading and writing.* London: Routledge & Kegan, Paul.

Taylor, D. (1983). *Young children learning to read and write.* Exeter, NH: Heinemann Educational.

Taylor, D. (1986). Creating family story: "Matthew! We're going to have a ride!" In W. H. Teale & E. Sulzby (Eds.), *Emergent literacy: Writing and reading* (pp. 139–155). Norwood, NJ: Ablex Publishing Corp.

Teale, W. H. (1986). Home background and young children's literacy development. In W. H. Teale & E. Sulzby (Eds.), *Emergent literacy: Writing and reading* (pp. 173–206). Norwood, NJ: Ablex Publishing Corp.

Tizard, B., & Hughes, M. (1984). *Young children learning*. Cambridge, MA: Harvard University Press.

Valentine, J., & Stark, E. (1979). The social context of parent involvement in Head Start. In E. Zigler & J. Valentine (Eds.), *Project Head Start: A legacy of the War on Poverty* (pp. 291–314). New York: Free Press.

Washington, V. (1988). Historical linkages between Black child development and social policy. In D. Slaughter (Ed.), *Black children and poverty: Developmental perspectives*. San Francisco: Jossey-Bass.

Washington, V., & Oyemade, V. (1987). *Project Head Start: Past, present, and future trends in the context of family needs*. New York: Garland Publishing.

White, B., & Watts, J. (1973). *Experience and environment: Major influences upon the development of young children*. New York: Prentice-Hall.

White, K. (1982). The relations between socioeconomic status and academic achievement. *Psychological Bulletin, 91,* 461–481.

Wilson, M. (1986). The Black extended family: An analytical consideration. *Developmental Psychology, 22,* 246–258.

Zigler, E., & Valentine, J. (Eds.). (1979). *Project Head Start: A legacy of the War on Poverty*. New York: Free Press.

Chapter 6

Literacy for American Indians: A Case Study

Sandra J. Fox
ORBIS Associates

There have been many articles in the literature about the special needs of ethnic groups in regard to English language skills. These articles stress the need to address cultural and language differences when planning and providing language instruction. An article by Fletcher (1983) summarizes literature which outlines problems American Indians may have when learning aspects of the English language. The article details problems which are probably due to language differences between English and native languages. The differences occur in phonology, morphology, syntax, and semantics. The article also points out that many Indian languages are highly complex systems. The use of "Indian English" by many Indian students is also noted. Reder (1988) stresses the need to address cultural factors which are beyond the more obvious culture and language differences when working with literacy development. He points out that cultural groups have "value systems and social meanings" specific to language functions and usage, and that these should be considered when planning literacy/language programs for various groups.

The literature has recommended the utilization of a language-experience approach to English literacy development for American Indian students (Fearn, 1971; Robinett, 1970; Simpson-Tyson, 1978). This approach is attuned to the language difference viewpoint, which suggests that when a student's English is different from standard

English it should not be viewed or treated as deficient. Because a viewpoint of language tends to characterize the viewpoint of culture as well, the holder of a language difference viewpoint also sees the student's culture as being different but not deficient. In this regard a language-experience approach utilizes the student's own language and experiences as the basis for instruction in literacy skills. Recent articles continue to recommend a language-experience approach for American Indian students (Lankford & Riley, 1986; Fox, 1988; Pearce & Gilliland, 1988).

American Indians have been the victims of many intervention programs designed to assimilate them into the dominant society in one way or another. The discussion to follow focuses on the history of these intervention attempts in regard to English literacy for the tribes of North and South Dakota, those tribes still served by the Bureau of Indian Affairs school system under the region called the Aberdeen Area. The tribes are various tribes of Sioux, the Turtle Mountain Chippewa (Ojibwa), and the Three Affiliated Tribes (Mandan, Hidatsa, and Arikara). The interventions will be described in terms of the language viewpoint or philosophy utilized in each: (a) the language deficit philosophy or viewpoint, which suggests that the language of some should be treated as deficient, with a focus on deficiencies; (b) the language difference viewpoint, as previously described; or (c) the no-difference philosophy, in which no special effort is made to accommodate language variance, whether it is viewed as deficient or different or not recognized at all.

In 1802, the Trade and Intercourse Act laid the basis for a Federal responsibility for the education of American Indians (Eder & Reyhner, 1986). In the second half of the 19th century, the emphasis of the Bureau of Indian Affairs education system was to "civilize" Indians in boarding schools, many of which were operated by churches through contracts with the government (U.S. Congress, 1969). Indian students from the Dakotas were sent to schools on the east coast to be removed from the influence of their families. The use of the students' native languages was strongly discouraged, and they were punished if caught speaking a native tongue. A language deficit viewpoint was definitely in place. This was also the time of the allotment period, when Indians on reservations were given allotments of land, 80 and 160 acres. One of the goals of the allotment (Dawes) act was to destroy tribal life by separating the people (Eder & Reyhner, 1986).

Around 1925, a movement of reform began in Indian education. A document called the Meriam Report called for change and denounced the allotment policy and the boarding school system. John Collier, Commissioner of Indian Affairs, implemented the building of more community day schools for Indians and promoted the training of teachers to recognize and meet the special needs of Indian students in a positive way. Some native language textbooks were written (Bureau of Indian Affairs, 1988). The language of the students was viewed from a language difference viewpoint, at least from the Washington Office of the Bureau. This movement was cut short, however, with the coming of World War II, and, therefore, this language program never came to fruition.

The time from 1945 to 1968 is called the *termination era*. This was the time of a relocation program designed to move Indians off of reservations and into urban areas for employment and assimilation purposes. The government also sought to cut off its special relationship with Indian tribes and to have all Indians be educated by public schools rather than maintain Bureau of Indian Affairs schools. Acts were passed that provided monies to public schools that educated Indian students (Eder & Reyhner, 1986). In the Aberdeen Area a special effort was made to have Bureau of Indian Affairs schools merge with public schools to form cooperative schools. The Bureau was again viewing the language of students from a deficit outlook, and the public schools in these situations brought to the Indian education scene a no-difference philosophy in regard to the language and literacy of Indian students, in that the materials and methods used for Indian students were the same as for non-Indian students.

From 1969 to the present, Indian education has been in an era termed one of *self-determination*. A major report which fueled this move was the Kennedy Report (*Indian Education: A National Tragedy)* of 1969 (Committee in Labor & Public Welfare). The idea is that Indian people are to have much more to say about the education of their children. School boards have been established for Bureau of Indian Affairs schools, and, with the passage of the Indian Self-Determination Act of 1975, Indian tribes can contract with the Bureau to operate their own schools (Bureau of Indian Affairs, 1988).

Where has this placed Indian education programs of the Aberdeen Area in regard to philosophy and viewpoint of language and literacy? Teacher training institutions in North and South Dakota have not emphasized a "different" language viewpoint in regard to the education of students who are different. Most teachers and administrators in Bureau or Bureau-contracted schools who were trained in North or South Dakota received training in the no-difference approach. Besides that, schools in the Aberdeen Area have tended to continue their programs based upon past experience, which has been in the deficit or no-difference approach. When queried about which approach they are using, they are likely to respond by telling you what textbook series they have.

Many of the children who enter Bureau schools in the Aberdeen Area have attended Head Start. The Head Start program has included some of the "different" philosophy in that some Indianness (aspects of Indian language and culture) has been included in the program but, in general, a no-difference instructional approach has been used, and sometimes Head Start staff have been known to try to teach the students too much, and emphasize skills to make up for the tradition of failure in the regular school program when compared to non-Indian school programs. The Follow-Through program implemented in the Aberdeen Area was the Pittsburgh Model, the Deficit Distar program of Bereiter and Englemann.

The intervention program which has perhaps had the largest influence upon the education program of the Bureau of Indian Affairs in the Aberdeen Area is Chapter I. Chapter I, by its very nature, has been from a deficit viewpoint because it focuses on students' deficiencies in relation to a standard norm. In the Aberdeen Area, Chapter I created individualized education for students in which they are tested with criterion referenced tests to determine deficiencies and provided with worksheets and other materials on various subskills of reading, language arts, or mathematics. The materials and methods used in the Chapter I program are "no different" from materials and methods used in Chapter I programs across the country. Most Chapter I programs in the Aberdeen Area schools have been pullout programs, thereby stigmatizing students and causing turf building between regular program and Chapter I teachers. The Bureau has been commended in the past by the United States Department of Education for its strict adherence to rules and regulations of the Chapter I program, but the overall effectiveness of the program has been disappointing.

Another program which has impacted the education of Indian students in Aberdeen Area schools has been Special Education of P.L. 94–142. The philosophy and approaches used by special education have been similar to those of Chapter I in that the emphasis is on noting deficiencies and "no-difference" materials and methods are utilized. A GAO report is critical of the Bureau for the many students who are labeled learning disabled (Native Research and Training Center, 1987).

The Title VII Bilingual Education program has been in place in some schools of the Aberdeen Area. When schools first received these programs, they were viewed as the way to include Indianness in the school system. Native and English languages were strengthened, and Indian culture instruction and materials were developed. The different philosophy or viewpoint was introduced through this program. The sad part is that Indian culture and language aspects have been a separate program from the regular school program, and that integration into the school curriculum has not taken place in the Aberdeen Area. Today, with Title VII's strong emphasis on improving English skills, it is often viewed as too similar to Chapter I to bother with.

The result of all of this is that 75 percent of the students in the Aberdeen Area are scoring below grade level on standardized tests. In 1985, 12th graders were scoring at the 9th-grade level in reading, math, and language arts (Fox, 1985). In 1986, 12th graders in North Dakota Bureau-funded schools were scoring at the 27th percentile, and 12th graders in South Dakota schools were scoring at the 22nd percentile (Bureau of Indian Affairs, 1988). Although the value of standardized tests can be questioned, especially for a minority population, Indian students are judged by them.

Where has all of this left Indian adults, parents, and grandparents of the children now in school? According to the 1980 census, 57 percent of Indian adults over 25 years old have not graduated from high school (Eder & Reyhner, 1986). The population of American Indians is a very young group compared to other groups. A labor market analysis of the United States Department of Labor and the Bureau of Indian Affairs states that the unemployment rates of Indians living on or near reservations is more than five times greater than the unemployment rate for all of the United States (Native Research and Training Center, 1987). Adult education programs on reservations in the Aberdeen Area are minimal; most of them are GED preparation programs.

How do Indian parents in the Aberdeen Area stand in relation to the education of their children? Parental involvement has only recently been encouraged in Indian schools. Indian parents, in general, surrender their children to what they view as an all-knowing school staff. They are quite sure that they have nothing to offer in regard to the educational programs. They expect the schools to be as they were when they attended, with use of textbook series, desks in straight rows, and so on. Many of the parents have been opposed to bilingual education, because they see it as possibly harmful to their children's progress in school. The emphasis on abstract, isolated subskills by the schools has made their children's school work even more foreign to them and, therefore, makes it more difficult for them to be interested or to assist their children with it. Further, schools are a major source of employment on reservations in the Aberdeen Area. Therefore, Indian teachers and aides are loyal to the school systems and their deficit and no-difference viewpoints.

The Bureau of Indian Affairs school system has continually strived to improve by providing for more local control and by reorganizing its administration. A review of what is taking place in instruction has been lacking. Presently the thrust of the Bureau is to utilize the

Effective School's Model (Bureau of Indian Affairs, 1988). Three of the characteristics of Effective Schools are:

1. Effective Schools have high expectations for student success. (Programs geared to the deficit and no-difference philosophies and viewpoints are not particularly geared toward having high expectations.)
2. Effective Schools have a clear sense of their educational mission and purpose. (Bureau schools need to establish a language/educational philosophy or viewpoint common to all programs.)
3. Effective Schools emphasize the development of a curriculum responsive to the student. (Bureau programs in the Aberdeen Area have been responsive to textbook curricula and federal program regulations.)

Hopefully, the Effective Schools intervention will help to turn Bureau schools around in regard to language development.

Some conclusions can be drawn from the history of intervention programs for Indian students in the Aberdeen Area. Furthermore, the conclusions can provide for recommendations in regard to intervention programs for any and all schools systems. These recommendations are:

1. Intervention programs, especially those which are legislated, should not be considered appropriate for all students.
2. Intervention programs should be based upon research into approaches recommended for specific populations being served.
3. Intervention programs should be implemented only with a full understanding of the philosophy of the instruction to be used and acceptance by the population to be served.
4. Legislated intervention programs must be flexible enough so that their own regulations do not make them counterproductive.
5. New intervention programs should not be added simply because others haven't done the job. Present programs need to be analyzed and changed so that they can provide the outcomes for which they were intended.

Intervention programs are necessary and good, but they need to be applied with much more care than is presently the case.

REFERENCES

Bureau of Indian Affairs. (1988). *Report on BIA Education.* Washington, DC: Author.

Eder, J., & Reyhner, J. (1986). The historical background of Indian education. In J. Reyhner (Ed.), *Teaching the Indian child*. Billings, MT: Eastern Montana College.

Fearn, L. (1971). Report of three pilot studies into initial reading: Navajo children. *Elementary English, 48,* 390–394.

Fletcher, J. D. (1983, October). What problems do American Indians have with English? *Journal of American Indian Education,* pp. 1–14.

Fox, S. J. (1988). A whole language approach to the communication skills. In H. Gilliland & J. Reyhner (Eds.), *Teaching the Native American*. Dubuque, IA: Kendall/Hunt Publishing Co.

Fox, S. J. (1985). *Aberdeen Area achievement study*. Unpublished paper for the Bureau of Indian Affairs.

Lankford, R., & Riley, J. D. (1986, May). Native American reading disability. *Journal of American Indian Education,* pp. 1–11.

Native Research and Training Center. (1987). *A study of the special problems and needs of American Indians with handicaps both on and off reservations*. Flagstaff, AZ: Northern Arizona University.

Pearce, D. L., & Gilliland, H. (1988). Developing reading skills. In H. Gilliland & J. Reyhner (Eds.), *Teaching the Native American* (pp. 113–120). Dubuque, IA: Kendall/Hunt Publishing Co.

Reder, S. M. (1987). Cooperative aspects of functional literacy development. In D. Wagner (Ed.), *The future of literacy in a changing world* (pp. 250–270). Oxford, UK: Pergamon Press.

Robinett, R. F. (1970, Spring). The teaching of English reading to American Indian children. *English for American Indians,* pp. 1–11.

Simpson-Tyson, A. K. (1978). Are Native American first graders ready to read? *The Reading Teacher, 31,* 798–801.

U.S. Congress. Senate. Committee on Labor and Public Welfare. (1969). *Indian Education: A national tragedy*. Senate Report, 91st Congress. Washington, DC: U.S. Government Printing Office.

Chapter 7

Parent Influences on Children's Mathematics

Herbert P. Ginsburg
Teachers College, Columbia University
Janine Bempechat
Harvard University
Y. Elsie Chung
Teachers College, Columbia University

Our ultimate goal is to develop programs enabling parents, particularly minority and poor parents, to help their children do well in school mathematics. Designing effective programs of this sort depends on an analysis of several issues, which we discuss in turn: the nature and causes of children's difficulties with mathematics, the ways in which parents influence children's cognitive and academic development, and the ways in which parents, including those who themselves have difficulty with school mathematics, can help children with their academic work.

MATHEMATICS ACHIEVEMENT: FACTS, NATURE, AND CAUSES

In this section we consider several aspects of school achievement in mathematics. First, we review the basic facts concerning American children's achievement; second, we present a cognitive analysis of their performance; and third, we discuss the causes of failure.

The Facts of Failure

The basic facts are that, as a group, American children do not do well in school arithmetic, and that minority and poor American children perform especially badly. Consider first cross-national studies comparing Americans and Asians. At all grade levels, American children display lower levels of academic achievement than do Asian children. For example, in a study of Japanese, Taiwanese, and American elementary school children, Stevenson, Lee, and Stigler (1986) found a consistent decline in mathematics achievement among the American children from kindergarten through fifth grade. These children were consistently surpassed by their Japanese and Taiwanese counterparts. Furthermore, a comparison of fifth-grade average mathematics achievement scores revealed that the *highest* U.S. score was lower than the *lowest* Japanese score. The average performance of the poorest U.S. *fifth*-grade class was only slightly better than the average performance of the best Japanese *first*-grade class.

At the high school level, the Second International Mathematics Study (McKnight et al., 1987) assessed mathematics achievement among eighth and twelfth graders in 20 industrialized nations. According to the SIMS study, at the eighth-grade level, U.S. students, while "slightly above the international average in computation" are "well below the international average in non-computational arithmetic" (McKnight et al., 1987, p. vi). At the twelfth-grade level, the very best U.S. students (highly selected to begin with, since so few U.S. students enroll in twelfth-grade mathematics) exhibit only average performance compared with foreign peers. The top one percent of Japanese students did better than the top one percent of students in all other countries; the top one percent of U.S. students had the *lowest* algebra scores of all countries and were among the lowest geometry scores; and the *average* Japanese outperformed the U.S. students in the top five percent in college preparatory mathematics. The clear finding is that, as a group, U.S. students perform relatively poorly on standard measures of mathematics achievement, and that, as a group, Asian students do especially well. An obvious caveat is that not *all* Asian groups do well. Thus, according to Lee (in press), Koreans living in Japan, where they occupy the status of an oppressed, caste-like minority, show relatively low levels of academic success.

The data also indicate that, within the U.S., minority and poor children, as groups, peform especially badly in school arithmetic. Data from the National Assessment of Educational Progress (Office of Education Research and Improvement, 1985–1986) reveal significant differences in mathematics achievement scores among ethnic groups, with white students performing better than minority groups. For example, in 1981–1982, the mean level of performance for all 13-year-olds in the United States was 60.5 on a standard achievement test. White children performed 2.6 points above this national average; their black and Hispanic counterparts scored 12.3 and 8.6 points, respectively, below the national norm. As students grow older, the gap between whites and minorities widens: minority students' scores decrease by approximately a further 2 points.

Similarly, the data show significant social class differences: at the age of 13, children's whose parents did not graduate high school (and hence, may be considered "lower class") scored 8.1 points below the national mean, whereas children whose parents attended college ("middle class") scored 3.2 points above the mean.

Not only do U.S. children achieve poorly in mathematics, they also dislike and avoid it. By the third grade or so, children often say they "hate" mathematics; they seem to experience great discomfort when they must produce number facts quickly and correctly; and some children feel that mathematics is the dullest of school subjects, about as boring and senseless as spelling. The situation seems to get even worse after childhood. In high school and college, many adolescents take as few mathematics courses as they can get away with, seldom studying topics beyond algebra, and in many cases arithmetic (McKnight et al., 1987). Indeed, a smaller proportion of U.S. students than of those in other countries receive exposure to advanced mathematical topics. And in the U.S., some individuals even suffer from "math anxiety," an extreme fear and loathing of mathematics.

U.S. students' attitudes towards mathematics are especially depressing. According to the SIMS study (McKnight et al., 1987), even highly selected twelfth grade students—that is, the select few who go on to study mathematics—believe that memorizing rules and formulas is basic to mathematics (p. 42); their attitudes toward mathematics are "rather neutral"; they find the subject "moderately important and somewhat easy"; "they do not particularly like it or dislike it" (p. 7).

We see, then, that American children as a group perform poorly in mathematics, and that minority and poor children do especially badly. Further, American children dislike mathematics, they avoid it, they see it as a subject requiring dull memorization, and even highly selected twelfth graders display no great enthusiasm towards learning it.

These findings raise two general sets of questions. The first refers to the nature of failure and success: What is involved in failure and what constitutes success? The second set of questions refers to the reasons for the failure and the success: Why do some children fail in mathematics, and why do Americans, particularly from poor and minority groups, perform poorly compared to Asians?

The Nature of Failure

The fact of failure is often obvious: a child gets low test scores, dislikes mathematics, and seems to understand little in the classroom. Yet recent cognitive research shows that matters are more complex than this. Several investigations (Ginsburg, 1989) have shown that children's mistakes are often generated by systematic error strategies or *bugs* (a term referring to systematic defects in computer programs). These are procedures that are defective in some way and that, in certain situations (but not necessarily all), result in regular, predictable errors. The student applies the bug quite conscientiously and accurately, but its fundamental flaw generates errors in certain problems. Some bugs are frequent and well known. For example, in subtraction one common bug occurs when the student operates according to the rule, "always take away the smaller from the larger." Applied to the problem 23 − 16, this bug results in the answer 13, because the student subtracts the three from the six, since "you always take the smaller from the larger." Sometimes this bug may lead to the correct answer, as in the case of 25 − 11, since borrowing is not an issue.

The notion of bugs asserts that wrong answers may not be random or meaningless. Children's mistakes often do not result from low *intelligence* or poor *mathematical aptitude* (whatever these terms may mean). The student makes mistakes for a reason. Moreover, the systematic procedures underlying children's errors usually have sensible origins in the child's learning history. For example, the rule "always subtract the smaller from the larger" may have originated in early subtraction lessons and in work sheet exercises; the student's mistake was to generalize it beyond its proper domain.

It is also clear that *cognitive schizophrenia* is a major phenomenon of mathematics learning. We use this phrase to refer to a split mind—that is to the commonly observed fact that, while many students possess a rather extensive and powerful informal knowledge of mathematics, they do not understand corresponding ideas in the areas of formal mathematics. For example, a child may add very well "in the head" but fail the same task when it is given in written form. Not only that, the child may see no relation between the two tasks. For example, one child insisted that the only way to determine the accuracy of her answer to a number fact question (she believed that eight times four is 35) was to ask someone. It was not possible, she felt, to solve the formal number fact problem by such informal procedures as counting objects or counting on the fingers, methods of which she was quite capable.

One way of interpreting cognitive schizophrenia is to say that children often seem to lose their natural intelligence when they get into school. They frequently regard mathematics as an arbitrary and meaningless exercise, a task to be endured only because the adult imposes it, a task without meaning in everyday life. They do their best with the algorithms taught (and in the process even devise sensible bugs), but they generally fail to relate what is taught in school to what they already know. Then this essentially meaningless learning of mathematics is combined with negative feelings about the subject and with faulty beliefs (for example, that mathematics is the subject in which one is required to get quick, accurate answers) to produce a constellation of failure. Moreover, this failure feeds upon itself. If arithmetic is meaningless to the young child, how much harder will it be for the older child to learn algebra? And if arithmetic is hard to learn, then positive attitudes toward it will be unlikely to develop.

Failure in mathematics is not necessarily limited to those who receive low achievement test scores. We should be skeptical about the meaning of success on standard achievement tests. We need to ask whether those scoring high on achievement tests are doing as well as they appear to. We have already seen that, according to the SIMS study, even highly selected and successful students believe that memorizing rules and formulas is basic to mathematics. Do our best students really understand mathematics, have a feel for it, approach it in a creative manner? Are there hidden costs in their success? Such questions remain to be answered, but answering them requires methods other than achievement tests. Everyone who has taken achievement tests, but not necessarily everyone who gives them, knows that they suffer from serious limitations. They tend to measure the conventional, the standard problem, the known formula; they do not necessarily measure anything very interesting or important from a mathematical point of view. High performance on such tests may not necessarily indicate a significant degree of mathematical talent, thought, or accomplishment. Since the tests provide such limited information, we should not let them define the goals of mathematics education. Indeed, some would argue that not only do the achievement tests fail to provide a solution, they are a significant part of the problem facing mathematics education. They constrict the conception of what mathematics is all about; they focus teachers and students on narrow aspects of performance. Teachers teach to the test and leave out the real mathematics.

So those who succeed on the standard tests may be employing a mechanical approach, may conceive of mathematics in narrow terms, and may be displaying little in the way of genuine mathematical creativity. Success of this sort is a form of failure, too. Whether or not this last point can be justified, the fact remains that many American children clearly do fail, and do not achieve real success. Consider next what real success involves.

The Nature of Success

One aspect of real success in mathematics involves *invented strategies*. Often students do not use the standard algorithms taught in school. Some students solve computational problems by means of invented strategies—that is, methods which are at least partially created by students themselves. These inventions typically draw on or modify what is taught in school and blend with what the child already knows. The results are often procedures which are remarkably effective and which children use with comfort and ease. A simple example involves early approaches to the number combinations. Given 3 plus 4, the child may not retrieve the number fact from memory as the teacher intends, but instead gets the result by reasoning from what is known ("It's 7, because I know that 4 plus 4 is 8, and this is just 1 less").

We can say that, in the most general sense, understanding mathematics is clearly more than using algorithmic procedures in an accurate way to get correct answers. It is more than the acquisition of skills. We believe (Ginsburg & Yamamoto, 1986) that an important aspect of understanding is the creation of links or connections among various areas of knowledge, formal and informal, procedural and conceptual. For example, one way to understand a written expression is to interpret it in terms of simpler informal procedures. Thus, 2 + 4 *must* be 6, because one can count up the parts to get the total. Another way to understand a written expression is connect it with events in everyday life. Thus, one boy we interviewed knew that 35 + 35 = 70, because he had bought packages of baseball cards at 35 cents each and knew from experience the cost of various numbers of packages.

We see then that success in school mathematics does not involve simply the mastery of basic skills. Indeed, an overemphasis on the traditional basic skills may prevent children from learning what is really important: to assimilate school mathematics into what they already know, to make school mathematics meaningful by connecting it with the informal and with other areas of knowledge. As we shall see below, effective programs of parent training must be based on valid notions of what children need to learn. Thus, just as the schools need to stress the cultivation of informal mathematics and the formation of connections, so do parents. Similarly, just as it is harmful for schools to place a narrow and rigid focus on the learning of basic skills, so parents should avoid this mistake too.

The Causes of Failure

Given our understanding of the nature of success and failure, we can now attempt to provide an explanation of the causes of failure Why do American children do so badly on school mathematics, and why do they as a group do worse than Asian students?

Cognitive inferiority. One possible explanation of Americans' low level of mathematics achievement is that they are less mathematically intelligent than are Asians. This is not a very good explanation. For one thing, research that Song and Ginsburg (1987) have conducted shows that Korean preschool children, 4 to 6 years of age, actually display a lower level of informal mathematical thinking than do their American counterparts. The Korean children do not begin school with some kind of intellectual head start. But once they are in school, the Korean children quickly surpass their American peers. This finding may be interpreted in terms of the effects of schooling and other cultural influences. Similarly, Stevenson, Stigler, Lee, Lucker, Kitmura, and Hsu (1984) found no differences in level, variability, and structure of basic cognitive skills among American, Japanese, and Taiwanese first and fifth graders. These cognitive skills were poor predictors of mathematics achievement in the different cultures. As we have already seen (Lee, in press), another bit of relevant evidence is that Koreans living in Japan, where they are an oppressed minority, similar to blacks in the United States, perform at a low level in school. The cause is clearly cultural; they did not abandon their natural intelligence upon immigration to Japan.

While it is unlikely that Asians are as a group more intelligent than Americans, or that they possess a greater degree of innate mathematical ability than do Americans, there may be some subtle cognitive reasons for their success in school mathematics. Miura, Kim, Chang, and Okamoto (in press) have recently shown that Asian children show a greater propensity than do Americans to spontaneously structure mathematical problems in base ten terms. This then can lead to a deeper understanding of arithmetic. While this gives Asian children an advantage, the effect is not powerful enough to explain the massive failure of American children at all ages.

Indeed, from a cognitive point of view alone, it is hard to understand why mathematics learning problems should be so prevalent. Cognitive development theories suggest at least two reasons for believing that people ought *not* to have an especially difficult time in learning mathematics. The first is that the vast majority of children doing poorly in arithmetic nevertheless possess the Piagetian thought structures, like concrete operations, which ought to be sufficient for understanding the subject matter (Allardice & Ginsburg, 1983). Indeed, as Piaget (1970) maintained, it is hard to understand why children have difficulty learning something what is in a sense merely an extension and formalization of what they already know.

We recognize of course that *some* individuals lack the basic intellectual equipment for learning mathematics. This must be true of many retarded persons. There may even be a few genuine cases of learning disability in mathematics (see, for example, McCloskey & Caramazza, 1985). But despite these exceptions it is abundantly clear that, for most individuals, *basic cognitive deficiency* is not responsible for learning difficulties in mathematics, at least at the level of the primary and secondary school curriculum. The child does not fail in mathematics because he or she lacks the concrete operations or has a faulty memory or a low IQ. Donaldson (1978) states this as a basic paradox: Why should the young child, so abundantly blessed with intellectual resources and so deeply committed to learning and exploration of the world, act so stupidly after a few years of school? "The problem then is to understand how something that begins so well can often end so badly" (p. 6).

A second reason for expecting people to learn mathematics with relative ease is that virtually everyone begins school with at least adequate, if not sophisticated, knowledge of *informal mathematics*. Children from preliterate societies in Africa (Saxe & Posner, 1983), street children in Brazil (Carraher, Carraher, & Schliemann, 1985), and lower-class children in the United States (Ginsburg and Russell, 1981) all demonstrate competence in basic concepts and procedures of arithmetic.

Our claim is *not* that there is a complete absence of social class differences in early mathematical thinking and experience. For example, Saxe, Guberman, and Gearhart (1987) have recently shown that, at the 4-year age level, working-class children score less well, on the average, than do middle-class children on relatively complex mathematical tasks. Rather, the claim is that, nonwithstanding these differences, children throughout the world exhibit a good degree of basic mathematical competence in the form of an informal mathematics.

Further, given this informal mathematics, children should not find it hard to learn what is taught in school. For example, even in preliterate cultures children spontaneously construct mental addition strategies embodying decomposition principles analogous to the base-ten system (Ginsburg, Posner, & Russell, 1981). Why should children capable of such sophisticated forms of mental addition experience difficulty learning the standard carrying algorithm?

Yet they do. Their problem is not that they lack an informal mathematics which could be used as a basis for understanding school mathematics, but that they fail to relate the two, and the result is cognitive schizophrenia. We argue, then, that learning difficulties in mathematics do *not* often result from a lack of basic cognitive ability, either in the sense of faulty intelligence, defective general intellectual skills like the concrete operations, or of a deficiency in informal mathematical knowledge. Learning difficulties certainly involve cognitive confusions and inadequacies, and even cognitive schizophrenia. But the main cause of all these problems is not basic cognitive deficiency. If so, then the basic question then is: What kind of factors interfere with our virtually universal intellectual abilities in such a way as to produce mathematics learning difficulties, especially in poor and minority American children? What are the causes of the massive ignorance, the fear and loathing of a key area of human knowledge?

Schooling. One obvious and serious explanation of learning problems refers to environmental impediments, specifically an inadequate educational system. Mathematics teaching is often unclear; textbooks may present materials in confusing ways; teachers may not convey the idea that mathematics *ought* to make sense; teachers may cover a variety of topics in a superficial manner, rather than present any single topic in adequate depth; class size may be too large; insufficient time may be devoted to mathematics instruction; and teachers suffer from math anxiety. Consider now some research which has been done on these matters.

Systematic study of *textbooks* in the United States, Japan, Taiwan, and the People's Republic of China (PRC) has revealed that mathematical concepts are introduced later in the U.S. than in these other countries (Bartasch & Ge, 1987; Stigler, Lee, Lucker, & Stevenson, 1982). Stigler et al. (1982) reported that many topics in Japanese elementary level texts are very advanced, not appearing in the U.S. texts until high school or college.

In this context, it is interesting to note that a recent study of the content of popular American mathematics texts showed that a relatively low percentage of new material is introduced from kindergarten through the fifth grade, and that the average percentage of new material decreases steadily from fifth to eighth grade (Flanders, 1987). Specifically, only 40 to 60 percent of material in second through fifth grade is new. Another way of looking at this is to say that new material is introduced only 2 to 3 days a week. By eighth grade, new material declines to 30 percent of the total, or new concepts are introduced only 1½ days per week. Each year, most new content is found in the second half of texts. Thus, at the beginning of the year, when students are fresh, they encounter old material; by the time new content is introduced, they are quite bored with mathematics. Researchers argue that this "spiraling" curriculum, in which much is reviewed and extended, may be undermining mathematics achievement (Flanders, 1987; McKnight et al., 1987). They point out that, in Japan, the magnitude of review is not nearly the same, and instruction builds on accumulated knowledge. Flanders concludes that students are trained to expect a low level of material from year to year. Good students are not challenged, and poor students learn that, if they do not master some material now, they will have the chance to catch up the following year.

Unfortunately, our students are lulled into a false sense of security, for the nature of mathematics instruction changes dramatically in ninth grade. With the introduction of algebra, courses and texts present three times as much new as old material (Flanders, 1987). As Flanders points out, the difficulty with algebra may have less to do with its inherent formality or unfamiliarity than with the fact that it is much more demanding relative to what students have been accustomed to for the previous 9 years.

American educators have argued that smaller *class size* might contribute to higher school achievement. In fact, Japanese class sizes are almost double those in the U.S. (Enloe & Lewin, 1987; McKnight et al., 1987; Stigler & Perry, 1987; Tobin, Wu, & Davidson, 1987). The average Japanese seventh grade has 41 students, compared to 26 students in the average American eighth-grade class (McKnight et al., 1987). Class size, then, need not be a factor in low U.S. mathematics achievement.

In discussions with Japanese and American preschool teachers, Tobin et al. (1987) found that large classes are used as a means to socialize groupism in Japan, a culture that emphasizes group relations over dyadic bonds. While Americans view the school as an extension of the home, with the teacher as a mother figure, the Japanese see it as a transition point between sheltered home life and the difficulties posed by the real world. The occasional chaos caused by large class sizes serves the purpose of preparing children for the realities of life. As we will see in a later section, this theme of preparing children for life's hardships is central in Japanese education (see White, 1987).

Consider next the *frequency and quality of mathematics instruction*. To begin, there is no evidence that Asian children get an earlier start in formal mathematics education. When kindergartens are compared, there is actually more formal instruction in American than in Asian classrooms (Ichikawa & Bacon, 1987; Stevenson, 1987b). Stevenson (1987a) has found that Asian mothers are concerned with their children's social experiences at the kindergarten level, while American mothers tend to focus on formal academic instruction. In this sample, 90 percent of U.S. mothers reported that they teach mathematics to their 5-year-olds, compared with 64 percent of Chinese mothers and 36 percent of Japanese mothers.

The *longer school year* in many Asian countries has led to the speculation that one of the reasons for Asian children's success is that they simply receive more mathematics instruction than their American counterparts. This may be the case in elementary school. The data show that, by the fifth grade, U.S. teachers spend three times as much time on reading compared to mathematics, while Japanese and Taiwanese teachers divide their instructional time about equally between both subjects (Stigler & Perry, 1987). By high school, however, McKnight et al. (1987) report no systematic relationship between length of school year and mathematics achievement, or between yearly hours of mathematics instruction and mathematics achievement. For example, the Japanese have the longest school year (243 days) and the highest mathematics achievement, but the Israelis, whose school year is just 17 days shorter, are in the bottom third in mathematics achievement. At the older end of the spectrum, the U.S. school year is 180 days long, and U.S. mathematics achievement ranks in the bottom third of nations. However, the Belgian school year is even shorter by 20 days, and their children rank in the top five nations in mathematics achievement. Instructional time ultimately does not appear to be an explanatory factor.

Stigler and Perry (1987) found that *mathematics instruction* is more coherent in Japan and Taiwan than in the States; that, in Japan, mathematics classes flow "like a good story." Japanese teachers tend to focus with some intensity on mathematical topics, explaining ideas in depth, and saving drill for homework. These classes stress reflection, verbalization, and self-criticism. According to Stigler and Perry (1987), "the word 'think' appears in our Japanese protocols more than twice as frequently as it does in either the Chinese or American protocols." By contrast, U.S. teaching flits around from topic to topic, does not treat anything in depth, engages students in senseless drill—senseless because the point of it either does not exist or is not understood—and in general does not allow students to make sense of the subject or even to believe that it *can* make sense. By contrast, it is common in the Asian countries to devote an entire class to understanding anywhere from one to three problems. In the U.S., as many as 40 problems can be addressed in one class period. The result is that American children, forced to cope with many segment changes during the course of one period, may not see what is the point of the lesson.

Stigler's research focused on elementary school. Can the high school be far behind? Though observations comparable to Stigler's do not seem to be available for the high school, the SIMS study shows that U.S. high school teachers stress rote learning of highly abstract material separated from concrete context, that they believe mathematics is easy to teach, that they blame poor student achievement on "lack of student ability" (p. xi), that they believe that "passive reception" is the most suitable technique for learning (p. 9), and that most of their classes involve lecturing at students.

In the U.S., there are often differences in the quality of education available to children of different social classes. Certainly the schools in many inner cities appear to be less adequate than those in the suburbs. Indeed, the inner city schools are perceived to be so inferior that the white middle class has abandoned them in favor of private schools.

Finally, many teachers, especially at the elementary level, themselves suffer from *math anxiety* and inadvertently transmit it to their students (Brush, 1981; Stodolsky, 1985).

In conclusion, it appears that Asian children may benefit from a more challenging, intense, and coherent curriculum than do Americans. This very curriculum may engender superior instruction. However, as we will see below, this more demanding curriculum is but one expression of a culture that sets no limits on what its members can attain in the way of intellectual knowledge and skills. Given the inadequacies of U.S. mathematics education, there should be little surprise that American students have trouble learning the subject. At the same time, deficiencies in instruction cannot fully explain individual differences in educational achievement and in feelings and attitudes toward the subject matter. In the same classroom, some children learn mathematics reasonably well, and others, despite adequate intellectual ability, learn it poorly; some like it and some hate it. Some individuals learn well even under poor instructional conditions, while others, despite adequate intellectual ability, fail to learn from good instruction. It has been noted that, in the U.S., Asian-American students often perform well in schools where others do poorly. Finally, a focus on instruction alone may not explain why some individuals who achieve well in the subject entertain such negative feelings towards the learning of mathematics. Since the nature of schooling cannot fully explain success or failure, we must look to other factors. In particular, we need to consider students' motivation, study skills, and the like, and the ways in which they are influenced by parents' world views and socialization practice.

COGNITIVE AND ACADEMIC SOCIALIZATION

We see then that, while all children—Asians, lower- and middle-class Americans—possess the intellectual capacity to do well in school mathematics, some of them fail. One contributor to the failure is poor education, for example in the form of incoherent mathematics lessons. Yet poor education alone does not explain everything. It does not entirely explain why some children are poorly motivated to learn and do not know how to go about learning in the school context. We now consider academic motivation and cognition, and the ways in which parents influence them.

The Distinction Between Cognitive and Academic Socialization

The question we have posed—what is known about how parents influence the motivation and school learning of their children?—is different from the way the question has sometimes been asked in the past. Some writers have framed the issue in terms of *cognitive socialization*. They have asked how parents affect the basic intellectual development of their children. Thus, some writers have proposed that lower class parents fail to provide their children with adequate stimulation and training, with the result that basic cognitive development is stunted, and this in turn produces school failure. Note that this "deficit model" focuses primarily, not on children's motivation or on their school learning, but on basic cognitive deficiencies that are presumed to produce school failure (for reviews, see Ginsburg, 1972, 1986).

In our view this is the wrong way to state the problem. It is wrong because, in general, school failure is not caused by basic cognitive deficiencies. Since most children failing in school, most poor children, most American children (when the contrast is with Asians), do not suffer from deficits in basic cognitive function (although they certainly suffer from confusions and other intellectual difficulties in connection with their knowledge of school subjects), it makes no sense to ask how parents produce such deficits. The more interesting question in connection with cognitive socialization deals with how parents facilitate basic cognitive development, if, indeed, they play an important role in promoting it.

But that is not our question either. Our question deals instead with *academic socialization,* not *cognitive socialization*. The chief issue is this: How do parents influence the development of skills, attitudes, and motives helpful for dealing with instruction in Academic schools, particularly in the area of mathematics? In some families, academic socialization operates in such a way as to produce a reasonable match between children's learning skills, attitudes, values, and motives, and the demands of the school. In other families, academic socialization operates in such a way that children have difficulty in capitalizing on their real potential, so that they fall behind in their school work and develop poor attitudes, expectancies, and motivations. Consider first some theories of academic socialization and then some research on Asian families which seem to be successful at academic socialization.

Theories of Academic and Cognitive Socialization

Theories of academic socialization vary along a dimension of generality. Theories focusing on general family characteristics associated with high achievement examine variables such as values and expectations. Early work in this area showed that children's general level of academic achievement is related to the degree to which parents pressure their children to succeed in school, reinforce the authority and prestige of the teacher, and provide tutoring when it is needed (Haggard, 1957; Rosen & D'Andrade, 1959; Toby, 1957).

More recently, researchers have found that the family variables positively associated with achievement include (a) mothers' success at predicting their child's performance on cognitive tasks (Miller, 1986); (b) mothers' strong press for achievement and satisfaction with past performance (Holloway, 1985); (c) parental beliefs that children actively construct their own reality (McGillicuddy-DeLisi, 1985; Sigel, 1985); (d) maternal appeals to reason, rather than to authority (McDevitt & Hess; 1985); and (e) high parental aspirations for children's educational or occupational attainment (Freeberg & Paine, 1967; Jones, 1972; Laosa, 1982; Marjoribanks, 1980).

In contrast to the approach to academic socialization, theories of cognitive socialization focus on the specifics of the interaction between children and parents. Some of this work has been heavily influenced by the writings of Vygotsky and Piaget, and assumes that parents function essentially as teachers. In this view, instruction may not be explicit, nor would it necessarily involve deliberate techniques or strategies. Rather, teaching is said to occur in many subtle and indirect ways, and is embedded in daily experience.

For example, Sigel (McGillicuddy-DeLisi, Flaughter, & Sigel, 1986; Sigel, 1982, 1985) has proposed that differences in parental "distancing" strategies may account for differences in cognitive development. *Distancing* is a construct that refers to the parent's attempt to help the child achieve a psychological separation from the immediate present. Distancing strategies vary along a continuum of *demandingness*. For example, observing and labelling are considered less demanding than proposing alternatives and resolving conflict. According to Sigel, parental distancing strategies facilitate children's representational thinking processes. Distancing strategies have been shown to be related to social factors and to child outcome measures. Empirical studies of middle- and working-class mothers have shown that middle-class mothers exhibit higher levels of questioning in a problem-solving task, and that middle-class children show higher levels of representational thinking than do working-class children.

In a related vein, Rogoff and her colleagues have proposed that the importance of the adult–child interaction process may lie in the guidance that adults provide in bridging the contexts of novel problems with more familiar ones (Rogoff & Gardner, 1984). According to Rogoff, adults "arrange the occurrence of cognitive tasks for children and facilitate the children's regulating the learning by difficulty of the task, providing well-placed pointers, and modeling mature performance" (p. 96). In this view, adults implicitly help children create a context in which new information becomes compatible with current knowledge and skill level.

Rogoff argues that the structuring of information serves as a scaffold for the learner, providing a framework in the search for a problem's solution. Learners can use the scaffold to support the performance of new aspects of a task that they may not have been able to handle alone. Rogoff notes that parents may not have explicit instructional goals, but may structure their interactions with their children in subtle ways that promote their children's social and cognitive development. In the learning process, information and skills are transmitted through pragmatic communication, and instruction occurs during the interaction. Initial instruction consists of highly supportive scaffolding, which eventually, through adult encouragement, gives way to greater participation by the child. Ideally, the child participates at a comforting but challenging level (the "zone of proximal development"), and the adult continually revises the scaffold for learning as the child's abilities develop. The adult who guides the child's growth in the most optimal way adjusts his or her support to levels just beyond what the child could manage alone.

While these theories of academic and cognitive socialization provide insight into what parents actually do to foster or inhibit achievement, the picture becomes much clearer when cultural factors are taken into account. In the next section, we examine cultural explanations for the apparent success of Asian children in mathematics. Specifically, we will show that a common Confucian ethic works to provide a cultural environment that fosters intellectual growth through effort and personal responsibility. These culturally based beliefs fuel an adherence to the value of discipline, the malleability of intelligence, high expectations and standards for performance, and realistic appraisals of the difficulty of mastering mathematics.

Empirical Studies

Cultural explanations: culture and family as academic socializers. Little is known about the academic socialization of mathematics achievement per se. Yet studies of socialization in Asian countries are at least indirectly relevant to the questions at hand. Researchers have suggested that a common Confucian Ethic—that individuals can always improve through effort and instruction—is an important factor in the achievement of Asians (Harrison, Serafica, & McAdoo, 1984; Nguyen & Henkin, 1983). Others have added that the traditional emphasis on respect for authority, obedience, perseverance, patience, restraint, and respect for the teacher as the bearer and provider of knowledge all weigh heavily in the school performance of Asians (Ayabe & Santo, 1972; Pitler, 1977; Sue, 1973; Sue & Kirk, 1972; Suzuki, 1980). Indeed, Suzuki (1980) and Vernon (1982) argue that the personality traits that contribute to success in school are the result of a socialization process that stresses, among other factors, respect for and obedience to authority, self-discipline, family and ancestral unity, and high-achievement motivation. Lee (1987) reports that, in response to the question "What is a good person?", the answers of Chinese parents reflected Confucian beliefs and included descriptors such as *moral, trusting, filial, responsible,* and *loves to learn.*

Cultural beliefs about the nature of the child also provide an important backdrop against which differential achievement patterns can be understood. In Japan, for example, the nature of the child is seen as inherently good (Kojima, 1986; Yamamura, 1986). In addition, there is a belief in universal equality of moral character and intellectual ability at birth. Any differences between children are seen as externally imposed; the possibility for change always exists. Therefore, the role of individual effort is paramount; success or failure is entirely within the child's control. White and LeVine (1986) report that a major goal in childrearing is to produce a child who will be "committed to and positively engaged in disciplined effort (White & LeVine, 1986, p. 57; see also Lebra, 1976; Vogel, 1971; White, 1987). Parents encourage the development of qualities that foster motivation in learning. These include (a) *gambaru,* or persistence; (b) *gaman suru,* or enduring hardship; (c) *sunao,* or compliance or obedience.

In fact, these qualities are fostered weeks before children begin formal schooling. Peak (1986) reports that the introduction to formal learning is carefully planned. It begins with a period of *minarai kikan,* or learning through watching. Children are invited to meet their teachers and older children weeks before they begin school. They observe older children in their classrooms and are introduced to their own classrooms during this period. After this introduction period, children are trained at home to follow certain basic routines that "form both the components of the skill to be learned and a cognitive support system for the learning process" (Peak, 1986, p. 117). Teachers encourage parents to train their children in daily study before they begin school. They may practice writing their names in a workbook. Daily concentration games and exercises train their ability to focus, and diligent care rather than speed of performance is reinforced.

With respect to discipline, it is interesting that 19th-century writings about child rearing stressed that children should not be overprotected (Kojima, 1986). Exposure to mild deprivation and hardship was seen as essential training in being able to face life's hardships in later adulthood. Today, enduring hardship is said to be important for the development of a mature individual. The hardship of difficult study, combined with sacrificing pleasure, is said to be good for the child (Befu, 1986; White, 1987; White & LeVine, 1986). According to Beauchamp (1987), there is a common view that the college entrance exams, widely criticized for the extreme pressure they place on students, are valuable, not for the information that becomes memorized, but for the intense, difficult, and lonely experience of preparation. This experience is believed to strengthen character and prepare students for life's difficult challenges. A similar attitude is apparent among Chinese parents. When asked what ideal characteristics children should have as adults, there was a high response rate for "can endure hardship" (Lee, 1987).

These beliefs are reflected in pedagogical practice today in Japan. Among the stated goals of the school syllabus is the belief that "it is desirable that, in the lower grades, one should learn to bear hardship, and in the middle grades, to persist to the end with patience, and in the upper grades, to be steadfast and to accomplish goals undaunted by obstacles or failures" (White, 1987, p. 17; see also U.S. Department of Education, 1987). In addition, the Japanese are vehemently opposed to ability grouping, reflecting their strong belief in the equality of innate ability (Enloe & Lewin, 1987; Holloway, Kashiwagi, Hess, & Azuma, 1986; White, 1987; White & LeVine, 1986). Classes are egalitarian in composition, as if all students are gifted and special (Enloe & Lewin, 1987).

How does this cultural philosophy get translated into parental practice? More specifically, how does it influence parents' beliefs about cognition and intelligence? We address this below.

The achievement beliefs of Asian vs. American parents. A review of the literature in this area leads one to conclude that Asian parents are very rich in what James Coleman calls "social capital"—

the norms, social networks, and relationships between adults and children that are important in children's growth (Coleman, 1987). Asian parents are very involved in, and responsible, with respect to their children's schooling (Enloe & Lewin, 1987; Uttal, Chen, & Diaz-Perez, 1987). In sharp contrast to the deteriorating fabric of American family life, Coleman believes that the high achievement of Asian children in this country reflects a strong family orientation towards academic achievement. The data appear to bear out this conclusion. Several cross-national studies of parental influence on children's learning have revealed differential patterns of effort attributions, expectancies and standards for performance, and involvement with schoolwork that may help to explain the differences in academic achievement that we see today (Hess, Azuma, Kashiwagi, Dickson, Nagano, Holloway, Miyake, Price, Hatano, & McDevitt, 1986b; Holloway et al., 1986; Kashiwagi, Azuma, Miyake, Nagano, Hess, & Holloway, 1984; Mordkowitz & Ginsburg, 1987; Stevenson, 1987a,b; Stevenson et al., 1986; Uttal et al., 1987). We address each of these factors below.

Attributions for success and failure. Given the Confucian emphasis on the malleability of ability, it is not surprising to find that Asian parents favor effort over ability as an explanation for their children's academic outcomes. What is interesting is the degree to which they do so relative to American parents. For example, Hess, Kashiwagi, Azuma, and Holloway (1986a) reported that Japanese mothers assigned more weight than American mothers to lack of effort as a cause of poor performance in mathematics. They were less likely than American mothers to blame the school, the child's ability, luck, or task difficulty for their child's poor performance. American mothers, while assigning the most weight to lack of effort, also emphasized lack of ability and poor instruction.

In a separate study, Hess and his colleagues found that Chinese-American and Caucasian-American mothers attempt to determine the reasons for their children's school failure, while mothers in the PRC punish their children for failure and become overtly angry with them (Hess, Chih-Mei, & McDevitt, 1987). This is interesting in light of Weiner's work on the link between affect and attributions. Bernard Weiner would argue that children whose mothers display anger over failure are likely to interpret that their mothers believe they can do better, and expect them to do so. This is more likely to motivate them towards greater effort in the future (see Weiner, Graham, Stern, & Lawson, 1982).

Hess concludes that there is a tendency among Japanese mothers to emphasize the role of internal, controllable factors in achievement, and that these beliefs get transmitted to their children. Japanese children believe effort to be effective and may be more likely than American children to respond to parental expectations. Stevenson adds that the apparent willingness of Chinese and Japanese children to work hard may be due to their parents' stronger belief in the value of hard work and the malleability of human nature (Stevenson et al., 1986).

Expectancies and standards for performance. Several researchers have found that, despite the higher achievement of their children, Asian parents are less satisfied with their children's school performance than are American parents. Stevenson and his colleagues found that 40 percent of the American mothers in their study were "very satisfied" with their children's school performance, as compared to less than six percent of the Chinese and Japanese mothers (Stevenson et al., 1986). Stevenson also found that U.S. mothers are much more satisfied than Chinese and Japanese mothers with the job that schools are doing—91 percent of American mothers, compared to 42 percent Chinese and 39 percent Japanese, felt the schools were doing an excellent job.

Uttal and his colleagues found that, for U.S. mothers to express dissatisfaction, their children had to be performing very badly, while mothers in the PRC expressed dissatisfaction with their children were doing only slightly poorly by comparison (Uttal et al., 1987). In this same study, both U.S. and PRC mothers believed that their children were performing at an above-average level. The Chinese mothers, however, had higher expectancies. They stated that they would be satisfied if their children scored 10 points *higher* than they expected them to. American mothers said that they would be satisfied if their children scored eight points *lower* than they expected them to. As we discuss at a later point, this kind of complacency may not foster the motivational tendencies necessary for high achievement.

Involvement with children's schoolwork. In their comparative study, Stevenson et al. (1986) found that Chinese and Japanese mothers were more involved than American mothers, in that more of them (a) reported that their children spent more time doing homework during the week and on weekends, (b) provided help on homework assignments, (c) provided a desk at home for their children's use, and (d) bought extra workbooks in mathematics and science. Similarly, a group of Asian American college students recently recalled that their parents were consistently concerned with, and supervised, their school progress (Mordkowitz & Ginsburg, 1987). Their recollections of their parents' behavior included strict control of after school hours, regulation of television viewing, and open discouragement of unlimited play.

The Achievement Beliefs of Asian and American Children

In light of the differences in Asian and American parents' achievement beliefs, it is interesting to ask whether their children adhere to similar beliefs. The data indicate that this is so. Hess and his colleagues have found that Asian children, like their parents, emphasize the role of internal, controllable factors in achievement, namely, effort (Hess et al., 1986a, b; Holloway et al., 1986). Furthermore, Japanese mothers and their children are more in agreement on this issue than American mothers and their children (Holloway et al., 1986). These researchers argue that Japanese children believe effort to be effective in improving school performance, and that, because of this, they may be more likely to respond to parental expectations.

Asian and American Children's Attitudes Toward Schooling

Given the different patterns of achievement-related beliefs held by Asian and American mothers, it is interesting to ask whether their children have different attitudes towards schooling. The literature reveals counterintuitive findings. For example, despite the fact that their school year is shorter and they have much less homework to do throughout the elementary school years, American children are less positive about school than Chinese children (52% American vs. 86% Chinese children "like school"—Stevenson et al., 1986).

With respect to mathematics and relative to their Asian peers, more American elementary school children think that it is "easy" or "very easy" (U.S.: 58% first grade, 40% fifth grade; Japan: 32% first, 30% fifth; Taiwan: 64% first, 22% fifth—Stevenson, 1987a).

At the high school level, where mathematics is more complex and sophisticated, it is remarkable to find that American students, whose mathematics achievement is among the lowest of 20 nations, like the subject and think that it is somewhat easy (McKnight et al., 1987). Japanese children, whose mathematics performance exceeds that of all other nations, believe that the subject is extremely difficult and admit to an extreme dislike for it. (As an aside, it is interesting that U.S. mathematics teachers think the subject is easy to teach, while their Japanese counterparts think that it is hard to teach.)

Furthermore, American children are also more optimistic about future performance than their Asian peers. Stevenson recently found that, among a sample of Chinese, Japanese, and American first and fifth graders, 75 percent of the American first graders expected that they would be the best in mathematics the following year, compared to 37 percent Japanese and 50 percent Chinese first graders (Stevenson,

1987a). A similar pattern was observed at fifth grade (58% American, 26% Japanese, and 29% Chinese expected to be above average or the best in mathematics in high school).

It is as if the Japanese students, through years of demanding curricula and hard work, develop a more realistic view of mathematics as a content area. The reality is that mathematics *is* very difficult and tedious in most schools, and becomes more so at increasingly advanced levels. The complacent attitudes of American mothers, teachers, and to some degree, textbooks, may ultimately be doing our children a serious disservice.

PARENT TRAINING PROGRAMS

Current Programs

There seem to be few programs designed to train parents to help their children with school mathematics. Hence we have reviewed parent programs, usually in the area of reading, to discover useful features and principles guiding their development. First we present the features and principles and then discuss an innovative example involving the use of video.

Supervision and reinforcement. Most programs stress parental supervision of homework and study, for example, by scheduling regular study hours. Programs also use various enrichment exercises, particularly during school breaks, to reinforce at home whatever was learned in school.

Role modeling. Programs encourage parents to serve as role models by reading at home in the presence of the child. Seeing the adult read is assumed to motivate the child and provide examples of the activity.

Social and affective component. Some programs encourage parents to convey the idea that learning is fun. Parents participate in reading activities with their children, and share the enjoyment.

Metacognitive element. Often following reading activities, parents are encouraged to engage the child in questions and discussions designed to foster an awareness of thinking and strategies.

Holistic learning experiences. Parents are encouraged to engage in activities which provide opportunities for learning through rich everyday experiences, like games or family trips.

Parent involvement. Programs vary in the extent to which they encourage active parent involvement. Some programs assign the parent to a passive role and essentially attempt to program his or her activities (we might call this *parent-proof education);* others use the parent as an active teacher and decision maker. Often both teachers and parents do not want parents to play too great a role in their children's learning. The jointly held assumption is that teacher knows best and needs to control the process of learning. Teachers sometimes see a threat in parents' providing "uncontrolled" (from their point of view) learning experiences outside of the classroom. Parent involvement requires a good amount of communication between parent and teacher and tends to be associated with individual attention to students' needs.

Use of video. The *Learning to Learn* program (Sprigle, 1974) offers several features of interest. One is that parents meet with teachers and teachers' assistants to view videotapes of large class and small group sessions. Viewing of the tapes helped parents to understand both what teachers did and how their own children learned. Discussion of the tapes was encouraged, so that parents could ask questions or contribute information concerning their own children. This was intended to prepare parents for conducting their own tutoring sessions at home, and also had the effect of reducing parents' feelings of intimidation with respect to the school and of establishing good working relations with the teachers.

Possible Programs

What should parent training programs involve and not involve in terms of content and method?

Beliefs, values, and expectations. We have seen that parental beliefs, values, and expectations concerning their children's school learning are central to academic success. Indeed, these intangibles seem to be more powerful than specific activities parents might undertake, like reading to their children. Thus, Asian-American parents who do not speak English well or at all, and who cannot help with homework in any specific way, can nevertheless transmit to their children basic values and expectations concerning education. This transmission is probably accomplished partly through direct exhortation, and partly through subtle forms of behavior.

If these values and expectations are so central, should parent programs attempt to convey them directly? The answer seems to be no, since they cannot be *taught* in a direct and simple manner. Beliefs develop in complicated ways from experience, cultural tradition, identification with others, and the like, not from superficial instruction. Furthermore, some beliefs depend on far more than experiences of the type that can be provided in parent-training sessions. No matter what they are told or experience in such settings, poor minority parents will not believe that education is important for jobs if they see a 40 percent employment rate around them. What will effectively influence their beliefs about this matter, and hence, the expectations that they convey to their children, is not lessons but real improvement in the economic prospects for minority groups and the poor. This, in turn, is a political and economic issue, not a question of training. It makes no sense to *train* parents to adopt certain beliefs about the value of education if there is in reality no basis for holding the beliefs.

Does this mean that we should ignore beliefs and expectations? No. Although we cannot teach beliefs, it may be possible to *influence* them or to help parents construct them by providing relevant experiences (of a type to be described below). In particular, parents can develop useful beliefs and expectations concerning their children's intellectual abilities.

Children's intellectual abilities. Poor and minority parents often hear from teachers and school psychologists that their children are intellectually deficient or otherwise responsible for their academic failure. It is now possible to give them a much different picture of their children's competence. Cognitive developmental psychology and modern educational psychology have a good deal to contribute in this regard, because they deal with important and valuable aspects of learning, not merely with achievement test scores. These disciplines show that poor children are indeed naturally intelligent, enjoy learning under certain circumstances, and can indeed do well in school. Poor parents need to know that their children are not intellectually inferior, not basically different from other children, and can learn effectively. There is a sound basis for the expectation that their children *could* do well in school under the right circumstances.

Video workshops. How can parents acquire such knowledge? One useful technique involves the *video workshop*, a method with which we have been experimenting for several years (Ginsburg, Kaplan, and Baroody, 1991). The essence of the method involves showing groups of individuals (teachers, parents, etc.) videotapes of children engaged in various intellectual tasks and being questioned, in an informal manner, concerning their methods or solution. For example, parents might see a 3-year-old engaged in a simple addition task involving concrete objects, of a 5-year-old explaining that the order of adding numbers makes no difference. Each video episode is dramatic, attention catching, and entirely typical of children at the age in question. As each video episode is presented, parents are asked to interpret it, to provide their own explanation of what it means. In this situation, the video episodes function as basic and compelling "data" concerning which "theories" can be constructed. The evidence provided by videos is direct, immediate, concrete, and hard to ignore. Our experience—so far with middle-class parents—shows that parents are captivated by the videos and develop from them important insights concerning their children's learning. These insights are determined partly by what we choose to put in the videotapes and partly by the interests of the parents.

Could such techniques work with poor parents? Our informal observations suggest that the prospects are good. Many poor parents seem to be vitally concerned with their children's schooling—at least at the outset—and indeed, one of the few academic subjects of interest to teenage mothers seems to be child development. An important challenge for the future is to develop for poor families meaningful programs of education concerning their children's intellectual capacities. Most likely these programs should center around video workshops rather than lectures or reading.

Empowerment in other skills. Yet knowledge about children is not enough. Programs should also empower parents in various ways. This is particularly necessary for the "invisible parents" of at-risk students. These parents often want their children to do well in school but feel powerless to help them, possibly because of their own difficult school histories or feelings of helplessness in controlling their current life situations. To bridge the gap between expectation and reality, between intent and action, programs must provide parents with the real possibility of helping their children.

One way this can be done is by helping parents to assume a more involved role in their children's learning, and to acquire specific teaching skills. Poor parents need to learn techniques for helping their children with homework and for dealing with teachers and schools. Some of the techniques will be very basic, like helping the child to budget time for homework and making sure it is completed, or like learning relevant questions to ask the teacher concerning the child's progress. It is also useful for parents to learn some of the substance of what children need to learn in mathematics, at least in the early years of school. Helping with homework demands some knowledge of the subject matter. But here we run up against the problem described at the outset of this chapter: in the U.S., poor children, who later become poor children's parents, do badly at mathematics and do not enjoy doing it. How then to get poor parents to understand basic mathematics so as to help their children with it? Our general hypothesis is that parents can learn some basic mathematics if it is presented in active and interesting ways—that is, if it is presented in the same ways it ought to be presented to children. Under these conditions it is possible for *anyone* to learn at least the rudiments of mathematics.

We have no evidence to support this conjecture, but can offer a relevant anecdote. We have been experimenting with a Japanese method of mathematics instruction involving concrete manipulatives appropriate for children from preschool to the 9th grade (Ginbayashi, 1984; Ginsburg & Yamamoto, 1986). In the course of presenting our preliminary results to teachers or interested parents, we show videotapes of children using the manipulatives to solve problems involving fractions. On watching the videos, parents and teachers get interested in the manipulatives, sometimes trying them themselves, and tell us that, for the first time, they now understand a topic like the multiplication of fractions.

Apparently, an active, manipulative approach can teach parents and teachers as well as children. The problem we think, will be to get parents to accept such an approach as legitimate, since they tend to believe, like most children who have been in school a while, that mathematics must be a rote, symbolic activity, as meaningless as it is painful. But through the active learning of mathematics, parents, like their children, could learn to stress the cultivation of informal mathematics and the formation of meaningful connections. We need research and development activities to determine the effectiveness of such parent education activities.

Finally, we note that parent education programs, no matter how well designed, are unlikely *by themselves* to make much of a difference unless they are embedded in other efforts at reform. Parents' ability to help their children with homework, for example, will count for little unless teachers also have ways of helping the children, and unless there is a real economic incentive for the poor children to do well in school.

ISSUES FOR THE FUTURE

The study of academic socialization is only in its infancy, and obviously much needs to be learned. Instead of offering a list—a very long one is possible—of issues that require investigation, we will focus on a few basic methodological issues. First, the generality of the conclusions deriving from the study of Asian parents is unknown. For example, Asian parents seem to exert strong control over their children's activities, and this seems to contribute to school achievement. Does this pattern characterize the academic socialization of other ethnic groups achieving academic success? Only research involving different cultures will provide general answers to questions of academic socialization.

Second, we require subtle measures. Simple questionnaires may provide superficial data concerning parental beliefs and practices. Intensive interviews and ethnographic observations (like Heath's, 1983) may be required making useful research a rather intensive undertaking. Furthermore, it may be useful to explore *indirect* methods, like projective tests, in which people do not simply say what they think ought to be said. Similarly, we require subtle measures of children's academic knowledge and motivation. The clinical interview technique should be used more than it is now to investigate children's knowledge of mathematics and other school subjects. And again, indirect methods for measuring children's beliefs and motives should be explored, as we are doing now (Mueller & Ginsburg, in press).

Third, we need to broaden our conceptions of the nature of academic motivation. While much research has focused on effort and ability attributions (thus showing that Asians seem to place greater emphasis on effort than do Americans), other and deeper aspects of motivation need to be considered, like the nature of one's ethnic identity and beliefs concerning caste status and economic prospects.

Finally, we require research and development involving parent education programs of various types. Can they really make a meaningful difference in the face of the massive failure of the educational system and the current economic prospects for poor and minority children?

REFERENCES

Allardice, B. S., & Ginsburg, H. P. (1983). Children's psychological difficulties in mathematics. In H. P. Ginsburg (Ed.) *The development of mathematical thinking* (pp. 319-350). NY: Academic Press.

Ayabe, H., & Santo, S. (1972). Conceptual tempo and the Oriental-American. *Journal of Psychology, 81*, 121-123.

Bartasch, K., & Ge, F. (1987, April). *Asian and American math textbooks and achievement patterns.* Paper presented at the biennial meeting of the Society for Research in Child Development, Baltimore.

Beauchamp, E. (1987). The development of Japanese educational policy, 1945–1985. *History of Education Quarterly, 27*, 299-324.

Befu, H. (1986). The social and cultural background of child development in Japan and the United States. In H. Stevenson, H. Azuma, & K. Hakuta (Eds.), *Child development and education in Japan.* New York: Freeman.

Brush, L. R. (1981). Some thoughts for teachers on mathematics anxiety. *Arithmetic Teacher, 29*, 37-39.

Carraher, T. N., Carraher, D. W., & Schliemann, A. S. (1985). Mathematics in the streets and in the schools. *British Journal of Developmental Psychology, 3*, 21-29.

Coleman, J. S. (1987). Families and schools. *Educational Researcher, 16*(6), 32-38.

Donaldson, M. (1978). *Children's minds*. London: Croom Helm.

Enloe, W., & Lewin, P. (1987). The cooperative spirit in Japanese primary education. *The Educational Forum, 51*, 233-237.

Flanders, J. (1987). How much of the content of mathematics textbooks is new? *Arithmetic Teacher, 35*(1), 18-23.

Flexer, J. F. (1986). The power of five: The step before the power of ten. *Arithmetic Teacher, 34* (3), 5-9.

Freeberg, N., & Paine, D. (1967). Dimensions of parental practice concerned with cognitive development in the preschool child. *Journal of Genetic Psychology, 111*, 245-261.

Ginbayashi, K. (1984). *Principles of mathematics education: Achievement of AMI*. Tokyo: The Association of Mathematical Instruction. (Available in English from The Association of Mathematical Interaction, 5-32-7, Hongo, Bunkyo-Ku, Tokyo 113, Japan).

Ginsburg, H. P. (1972). *The myth of the deprived child: poor children's intellect and education*. Englewood Cliffs, NJ: Prentice-Hall.

Ginsburg, H. P. (1986). The myth of the deprived child: new thoughts on poor children. In U. Neisser (Ed.), *The school achievement of minority children: new perspectives*. Hillsdale, NJ: Erlbaum.

Ginsburg, H. P. (1989). *Children's arithmetic: A psychological approach to mathematics education* (2nd ed.). Austin, TX: Pro Ed.

Ginsburg, H. P., Kaplan, R. G., & Baroody, A. J. (1991). *Children's mathematical thinking: Video tape workshops for educators*. Evanston, IL: Everyday Learning Corporation.

Ginsburg, H. P., Posner, J. K., & Russell, R. L. (1981). The development of mental addition as a function of schooling and culture. *Journal of cross-cultural psychology, 12*, 163-178.

Ginsburg, H. P., & Russell, R. L. (1981). Social class and racial influences on early mathematical thinking. *Monographs of the society for research in child development, 46*, 16 (Serial No. 193).

Ginsburg, H. P., & Yamamoto, T. A. (1986). Understanding, motivation, and teaching: comment on Lampert's "Knowing, doing, and teaching multiplication." *Cognition and Instruction, 3*(4), 357-370.

Haggard, E. (1957, Winter). Socialization, personality, and academic achievement in gifted children. *The School Review, 65*, 388-414.

Harrison, A., Serafica, F., & McAdoo, H. (1984). Ethnic families of color. In R. D. Parke (Ed.), *Review of child development research* (Vol. 7, pp. 329-371). Chicago: University of Chicago Press.

Heath, S. B. (1983). *Ways with words: Language, life and work in communities and classrooms*. Cambridge, UK: Cambridge University Press.

Hess, R., Chih-Mei, C., & McDevitt, T. (1987). Cultural variations in family beliefs about performance in mathematics: Comparisons among the People's Republic of China, Chinese-American, and Caucasian-American families. *Journal of Educational Psychology, 79*, 179-188.

Hess, R., Kashiwagi, K., Azuma, H., & Holloway, S. (1986a). *Causal attributions by Japanese and American mothers and children about performance in mathematics.* Unpublished manuscript.

Hess, R., Azuma, H., Kashiwagi, K., Dickson, W., Nagano, S., Holloway, S., Miyake, K., Price, G., Hatano, G., & McDevitt, T. (1986b). Family influences on school readiness and achievement in Japan and the United States: An overview of a longitudinal study. In H. Stevenson, H. Azuma, K. Hakuta (Eds.), *Child development and education in Japan.* New York: Freeman.

Holloway, S. (1985). *The relationship of mothers' beliefs to children's mathematics achievement: Some effects of child gender.* Unpublished manuscript.

Holloway, S., Kashiwagi, K., Hess, R., & Azuma, H. (1986). Causal attributions by Japanese and American mothers and children about performance in mathematics. *International Journal of Psychology, 21*, 269-286.

Ichikawa, V., & Bacon, W. (1987). *Parental expectation and kindergarteners' scholastic achievement in the United States and Japan.* Paper presented at the biennial meeting of the Society for Research in Child Development, Baltimore.

Jones, P. (1972). Home environment and the development of verbal ability. *Child Development, 43*, 1081-1086.

Kashiwagi, K., Azuma, H., Miyake, K., Nagano, S., Hess, R., & Holloway, S. (1984). Japan-U.S. comparative study on early maternal influences upon cognitive development: A follow-up study. *Japanese Psychological Research, 26*, 89-92.

Kojima, H. (1986). Child rearing concepts as a belief-value system of the society and the individual. In H. Stevenson, H. Azuma, & K. Hakuta (Eds.), *Child development and education in Japan.* New York: Freeman.

Laosa, L. (1982). Families as facilitators of children's intellectual development at 3 years of age: A causal analysis. In L. Laosa & I. Sigel (Eds.), *Families as learning environments for children* (pp. 1-45). New York: Plenum.

Lebra, T. (1976). *Japanese patterns of behavior.* Honolulu, HI: University Press of Hawaii.

Lee, L. C. (1987). *China's integrated approach to child care and socialization.* Invited address, SRCD, Baltimore.

Lee, Y. (in press). Koreans in Japan and the United States. In M. Gibson & J. Ogbu (Eds.), *Minority status and schooling: A comparative study of immigrant and involuntary minorities.* New York: Garland.

Majoribanks, K. (1980). *Ethnic families and children's achievements.* Sydney: Allen & Unwin.

McCloskey, M., & Caramazza, A. (1985). Cognitive mechanisms in number processing and calculating: Evidence from dyscalculia. *Brain and Cognition, 4*, 171-196.

McDevitt, T., & Hess, R. (1985, April). *Children's beliefs as a route of influence between mothers' direct control techniques and children's achievement in mathematics*. Paper presented at the Biennial Meeting of the Society for Research in Child Development.

McGillicuddy-DeLisi, A. (1985). The relationship between parental beliefs and children's cognitive level. In I. Sigel (Ed.), *Parental belief systems* (pp. 7-24). Hillsdale, NJ: Erlbaum.

McGillicuddy-DeLisi, A., Flaugher, J., & Sigel, I. (1986). Familial influences on planning. In J. Kagan (Ed.), *Blueprints for thinking* (pp. 395-427). Cambridge, UK: Cambridge University Press.

McKnight, C. C., Crosswhite, F. J., Dossey, J. A., Kifer, E., Swafford, J. O., Travers, K. J., & Cooney, T. J. (1987). *The underachieving curriculum: Assessing U.S. school mathematics from an international perspective*. Champaign, IL: Stipes Publishing Co.

Miller, S. (1986). Parents' beliefs about their children's cognitive development. *Child Development, 59*, 259-285.

Miura, I. T., Kim, C. C., Chang, C. M., & Okamoto, Y. (in press). Effects of language characteristics on children's cognitive representation of number: Cross-national comparisons. *Child Development*.

Mordkowitz, E. R., & Ginsburg, H. P. (1987). Early academic socialization of successful Asian-American college students. *Quarterly Newsletter of the Laboratory for Comparative Human Cognition, 9*, 85-91.

Mueller, E. C., & Ginsburg, H. P. (in press). *The Mueller-Ginsburg projective test (MUG)*.

Nguyen, L. T., & Henkin, A. (1983). Perceived sociocultural change among Indochinese refugees: Implications for education. In R. Samuda & S. Woods (Eds.), *Perspectives in immigrant and minority education*. New York: University Press of America.

Office of Educational Research and Improvement. (1985-1986). *Digest of Education Statistics (1985-1986)*. Washington, DC: U.S. Department of Education, Center for Statistics.

Peak, L. (1986). Training learning skills and attitudes in Japanese early educational settings. In W. Fowler (Ed.), *Early experience and the development of competence* (Vol. 38, pp. 111-123) (New Directions for Child Development). San Francisco: Jossey-Bass.

Piaget, J. (1970). *Science of education and the psychology of the child*. New York: Orion Press.

Pitler, B. (1977). Chicago's Korean American community. *Integrated Education, 88*, 44-47.

Rogoff, B., & Gardner, W. (1984). Adult guidance of everyday cognition. In B. Rogoff & J. Lave (Eds.), *Everyday cognition: Its development in social context*. Cambridge, MA: Harvard University Press.

Rosen, B., & D'Andrade, R. (1959). The psychosocial origins of achievement motivation. *Sociometry, 22*, 185-218.

Saxe, G. B., Guberman, S. R., & Gearhart, M. (1987). Social processes in early number development. *Monographs of the Society for Research in Child Development, 52*(Serial number 216).

Saxe, G. B., & Posner, J. (1983). The development of numerical cognition:

Cross-cultural perspectives. In R. B. Davis (Ed.), *The development of mathematical thinking* (pp. 291-317). New York: Academic Press.
Sigel, I. (1982). The relationship between parental distancing strategies and the child's cognitive behavior. In L. Laosa & I. Sigel (Eds.), *Families as learning environments for children*. New York: Plenum Press.
Sigel, I. (1985). A conceptual analysis of beliefs. In I. Sigel (Ed.), *Parental belief systems* (pp. 345-371). Hillsdale, NJ: Erlbaum.
Song, M. J., & Ginsburg, H. P. (1987). The development of informal and formal mathematical thinking in Korean and American children. *Child Development, 58*, 1286-1296.
Sprigle, H. (1974). Learning to Learn program. In S. Ryan (Ed.), *A report on longitudinal evaluations of preschool programs* (Vol. 1, pp. 109-124). Washington, DC: Department of Health, Education, and Welfare.
Stevenson, H. (1987a). The Asian advantage: The case of mathematics. *American Educator, 47*, 26-31.
Stevenson, H. W. (1987b). American's math problems. *Educational Leadership, 45*(2), 4-10.
Stevenson, H., Stigler, J., Lee, S., Lucker, G., Kitmura, S., & Hsu, C. (1984). Cognitive performance and academic achievement of Japanese, Chinese, and American children. *Child Development, 56*, 718-734.
Stevenson, H., Lee, S. S., & Stigler, J. (1986). The mathematics achievement of Chinese, Japanese, and American children. *Science, 56*, 693-699.
Stigler, J., Lee, S., Lucker, G., & Stevenson, H. (1982). Curriculum and achievement in mathematics: A study of elementary school children in Japan, Taiwan, and the United States. *Journal of Educational Psychology, 72*, 315-322.
Stigler, J. W., & Perry, M. (1987). *Cross-cultural studies of mathematics teaching and learning: Recent findings and new directions*. Paper delivered at the NSF-NCTM Research Agenda Conference, Columbia, MO.
Stodolsky, S. S. (1985). Telling math: Origins of math aversion and anxiety. *Educational Psychology, 20*, 125-133.
Sue, D. (1973). Ethnic identity: The impact of two cultures on the psychological development of Asians in America. In S. Sue & N. Wagner (Eds.), *Asian Americans: Psychological perspectives*. Lomond, CA: Science and Behavior Books.
Sue, D., & Kirk, B. (1972). Psychological characteristics of Chinese-American students. *Journal of Counseling Psychology, 19*, 471-478.
Suzuki, B. (1980). Education and the Socialization of Asian-Americans: A revisionist analysis of the "Model Minority" thesis. In R. Endo, S. Sue, & N. Wagner (Eds.), *Asian-Americans: Social and psychological perspectives, II*. New York: Science and Behavior Books.
Tobin, J., Wu, D., & Davidson, D. (1987). Class size and student-teacher ratios in the Japanese preschool. *Comparative Education Review, 31*, 533-549.
Toby, J. (1957). Orientation to education as a factor in the school maladjustment of lower class children. *Social Forces, 38*, 259-266.
U. S. Department of Education. (1987). *Japanese education today* (Rep. No. OR87-500). Washington, DC: U.S. Government Printing Office.

Uttal, D. C., Chen, C., & Diaz-Perez, R. (1987). *Parental beliefs and children's school achievement in the United States and China*. Paper presented at SRCD, Baltimore.

Vernon, P. (1982). *The abilities and achievements of Orientals in North America*. New York: Academic Press.

Vogel, E. (1971). *Japan's new middle class*. Berkeley, CA: University of California Press.

Weiner, B., Graham S., Stern, P., & Lawson, M. (1982). Using affective cues to ensure causal thoughts. *Developmental Psychology, 18*, 278-286.

White, M. (1987). *The Japanese educational challenge: A commitment to children*. New York: Academic Press.

White, M., & LeVine, R. (1986). What is an "ii ko" (good child)? In H. Stevenson, H. Azuma, & K. Hakuta (Eds.), *Child development and education in Japan*. New York: Freeman.

Yamamura, Y. (1986). The child in Japanese society. In H. Stevenson, H. Azuma, & K. Hakuta (Eds.), *Child development and education in Japan*. New York: Freeman.

Chapter 8

What Mathematical Cognitive Skills Do Parents Offer Children?

Sandra P. Marshall
San Diego State University

Before we address the issue of how to facilitate parents' transferral of mathematical cognitive skills to their children, we need to identify which skills are sufficiently important to warrant the effort. Clearly, we don't want to waste time developing skills that have little value. Identification of skills transmitted by parents of mathematically literate children is in order.

One way to make that identification is to examine differences in parent–child interactions among cultural groups. This approach is taken by Ginsburg and Bempechat. They compare values, beliefs, and attitudes of Asian parents, whose children are demonstrably successful in mathematics, with those of American parents, whose children have more difficulty.

As Ginsburg and Bempechat point out, parents of students in the United States have different attitudes and take different actions towards the learning of mathematics than do parents of students in Taiwan and Japan. It isn't just the case that Asian parents value schooling more than American parents (although this may be true). What is important in the Ginsburg and Bempechat review is the description of the activities of parents with respect to their children's learning. Asian parents are directly involved in their children's mathematics learning. They work with their children, monitor their homework, talk with their teachers, and set high standards for achievement. In so doing, they apparently transmit more than knowledge about particular problems and situations in mathematics; we infer that they also transmit a body of cognitive skills and beliefs that govern their own attitudes towards mathematics and help to shape their children's attitudes.

As we investigate the intergenerational transfer of cognitive skills in mathematics, it is necessary for us to ask: Which skills are important for parents (rather than teachers) to transmit to their children? Certainly we don't expect parents to be surrogate teachers, drilling students on concepts, algorithms and number facts. What do we expect? And, if we can make explicit our expectation, how can we convince parents of its importance and value?

Ginsburg and Bempechat make an important beginning in answering these questions. They summarized many differences between Asian and American parents in terms of expectations, attributions, and beliefs about students' performance. As they point out, we have few cross-cultural studies about parental influence in mathematics learning. We need to infer particular strategies about acquiring cognitive skills in mathematics from broad comparisons about schooling in general.

We also have few intervention programs that focus on parental participation in the acquisition of children's mathematical knowledge. Again, we must use studies of other areas—particularly in reading—to hypothesize the contents of mathematics programs.

What would an intervention program for fostering children's cognitive skills in mathematics look like? Ginsburg and Bempechat suggest some important principles. I think four of the most important are the following:

1. Parents need to supervise students' study.
2. Parents need to be role models in using important skills such as persistence and attribution to external sources.
3. Parents need to convey the value and pleasure of learning and using mathematics.
4. Parents and children need to work together in activities that use various mathematics skills.

These are valuable guides, and the remainder of this chapter will discuss and expand various implications of these four principles.

Supervision. When we speak of parents supervising their children's studying and homework, we mean several things. First, we expect parents to establish a routine and schedule for doing the homework. It is important that this routine be jointly observed by parents and children; doing homework is not punishment and should not correspond to banishment from the family circle.

Supervision also means active involvement as necessary on the part of the supervisor. Parents are not expected to answer every question and solve every problem contained in the homework, but they should be prepared to do so and willing to explain various concepts to the student. Even if parents cannot solve the problems themselves, it is probably valuable to students if parent and child can sit together and discuss the problems or the text that is not understood.

A major difficulty is that many parents themselves do not like mathematics and do not know how to solve certain problems. In training parents to be supervisors, we want to train them also to be facilitators and to help them work with their children productively. We cannot expect them to be mathematics teachers and to perform the same tasks that classroom teachers carry out. Homework tasks must be carefully designed to elicit cooperation and interaction between parent and child. Recognition of the parent's role in homework is now acknowledged, and ways of facilitating that role are being developed (cf. California State Department of Education, 1987, p.3).

Role models. A large body of psychological literature exists about how children learn from the observation of role models. Currently, many reforms in mathematics education focus on the need for the teacher to model different strategies in solving problems and to model various problem-solving activities. It is equally important for parents to provide role models in using mathematics, but many parents don't know how to do this.

Most of our information about parental role modeling comes from studies of reading. Children observe their parents reading and seek to imitate them. It is hypothesized that children of parents who read frequently will themselves be better readers and will enjoy reading more than children whose parents do not read.

If role models in reading are so effective, can we make a parallel for children who observe their parents using mathematics? What would the role model do? Certainly we do not envision adults working on pages of multiplication problems. One form of role model is found in the apprentice–master relationship, described by Lave, Smith, and Butler (1988). They document how Liberian children learn tailoring skills (including mathematical ones) through apprenticeship. It is clear that the skills they learn go well beyond ordinary classroom procedures. Lave's study suggests that we may profitably study adults and examine the skills they use in daily life. Children, as "apprentices," may then imitate those skills and develop appropriate cognitive abilities.

One difficulty for parents is that they frequently are unaware of their own uses of mathematical skills. Too often adults (and children) equate mathematics with arithmetic computation. A valuable product of an intervention program would be the recognition of several problems which parents routinely encounter that have identifiable mathematical content, that are familiar situations for parents, and that could be easily conveyed to children.

For example, in planning a grocery trip, one might consider how many items to buy, whether one can carry all of the items once purchased, whether one already has some necessary ingredients, and how much more is needed. Children can learn to use information about cost, weight, and comparisons, and they can see the value of making estimates. Problems such as this demonstrate that it isn't necessary to carry out arithmetic computations to use mathematical knowledge and reasoning.

Some of the skills used by parents in solving problems like the grocery trip may be termed *intuitive mathematics*. Individuals may know how much they can carry by having a memory of the size and shape of various products. They may not need to compute the volume of a shopping bag or convert ounces to pounds. What they already know is that they need to attend to the important dimensions and to combine information about those dimensions (e.g., weight of the items is more important than color when you are trying to decide how much you can carry for three blocks). This is valuable knowledge to share with children.

If parents are apprehensive about working with children because the parents have weak computational skills, we need to find areas of intuitive mathematics in which the parents feel competent. We cannot expect parents to teach all concepts of mathematics. It will be a big step if we can help them identify ways in which they themselves use and value mathematical knowledge. What we want of the parents is for them to model for their children how mathematical knowledge enters their own lives.

Value of mathematical knowledge. Ideally, the value of mathematical knowledge will be obvious as parents realize for themselves how much mathematical knowledge they do have and how much they use it. Without an understanding of its value, parents and students are likely to view the learning of mathematics as a punishment, a barrier, or a meaningless activity.

The use of mathematical skills should be empowering to parents and children alike. Many educators are aware of the importance of showing individuals that having particular kinds of knowledge gives them personally more power over their own lives. This is a fundamental issue that all intervention programs have to address if they are to be successful. Students and parents alike drop out of special programs if they fail to see personal advantage.

We want parents and children to value mathematical reasoning skills. We also want them to like using these skills. Affect has a strong influence on children's problem-solving success (Marshall, 1989). Many parents and teachers apparently have negative attitudes towards mathematics (Ernest, 1976). Successful intervention programs can address directly the issue of parental attitude toward and valuation of mathematical cognitive skills.

Family activities. As others have pointed out (cf. Rogoff & Gardner, 1984), parents can transmit a variety of cognitive skills to their children through carefully planned problem-solving tasks. What is important in these tasks is that parents and children share the responsibility of problem solving and work together to reach a solution.

One of the most important cognitive skills that parents can transmit to their children is the ability to work on problems for an extended period of time. In too many mathematics classrooms, children learn that mathematics problems have simple form and can be solved in a matter of minutes. Few problems are presented to them for which they have the opportunity to think about the situation and to engage in real problem solving (cf. Schoenfeld, 1988). Yet real problems are rarely conceived and solved in a few minutes.

Children (and parents) need to become cognizant of the intricacies of real-world problems involving mathematics. For example, consider the practical problem of installing a lawn sprinkler system. There are many dimensions of the problem to be considered, and typically the installer makes several approaches before buying the needed pipes and sprinkler heads. Similarly, when one wants to build a stereo cabinet, there are a number of decisions to be made. The builder must decide how tall to make the cabinet, how many shelves to build, how heavy the finished cabinet will be, and so on. Parents and children can work together on many real projects such as these. Parents can share some of the decision-making processes with children, and both can observe how and where mathematical reasoning is needed.

Many successful teachers of mathematics classes in elementary school already use this technique. Consider the teacher, for example, who has her class decide how large the playground is and whether another tetherball court could be added. Or the one who has her pupils plan the class party at the end of the year, deciding on the amount of refreshments, cups, plates, and so on, to be purchased. These are problems that involve mathematical reasoning together with knowledge of the world. These problems take more than a few minutes to resolve. And they have more than one solution.

The four principles discussed here are not the only ones that will guide a successful intervention program for parents. As Ginsburg and Bempechat note, there are others that will be important. Nonetheless, these four provide an essential starting point from which we can generate strong programs to assist parents in recognizing and transferring mathematical cognitive skills to their children.

REFERENCES

California State Department of Education. (1987). *The mathematics model curriculum guide, kindergarten through eighth grade.* Sacramento: California State Department of Education.

Ernest, J. (1976). Mathematics and sex. *American Mathematics Monthly, 83,* 595-614.

Lave, J., Smith, S., & Butler, M. (1988). Problem solving as an everyday practice. In R. I. Charles & E. A. Silver (Eds.), *Research agenda in mathematics education: The teaching and assessing of mathematical problem solving* (pp. 61-81). Reston, VA: National Council of Teachers of Mathematics, and Hillsdale, NJ: Erlbaum.

Marshall, S. P. (1989). Affect in schema knowledge: Source and impact. In D. B. McLeod & V. M. Adams (Eds.), *Affect and mathematical problem solving: A new perspective* (pp. 49-58). New York: Springer-Verlag.

Rogoff, B., & Gardner, W. (1984). Adult guidance of cognitive development. In B. Rogoff & J. Lave (Eds.), *Everyday cognition: Its development in social context.* Cambridge, MA: Harvard University Press.

Schoenfeld, A. (1988). When good teaching leads to bad results: The disasters of "well-taught" mathematics courses. *Educational Psychologist, 23,* 145-166.

Chapter 9

The Higher-Order Cognitive Skills: What They Are and How they Might be Transmitted[1]

Susan F. Chipman
U.S. Office of Naval Research

We seem quite comfortable with the concept of "higher-order cognitive skills" as long as we remain in casual discourse, where vague and nebulous ideas are acceptable. But concepts that are to guide teaching and testing, or even a discussion of the ways in which higher cognitive skills might be transmitted from one generation to the next, require a more solid foundation. It is not enough merely to say that these skills are characterized by all sorts of positive-sounding qualities: a non-algorithmic character, complexity, multiple possible outcomes, application of multiple criteria, evidence of nuanced judgement, and so on (Resnick, 1987). Behind our choice of the term *higher order*, there are strong intuitions about the way in which our cognitive activities are structured and controlled. In this chapter, I explore those intuitions in

---

[1] The initial section of this chapter dealing with the definition of higher-order cognitive skills and the cognitive science view of human cognition closely resembles a paper that was written in September 1986, to provide advice to the Blue Ribbon Panel examining The National Assessment of Educational Progress. That paper, "What is meant by 'Higher-order cognitive skills,' is available from the ERIC system as ED 279 668 or TM 870 055, and includes many references to implications for testing that have been omitted in this version. This chapter expresses the personal professional opinions of the author and does not necessarily represent the official policy or position of the Office of Naval Research or of any other agency of the U.S. Government.

some depth, linking them to recent research and theory in cognitive science that gives them scientific meaning and a clearer definition.

The higher-order cognitive skills are those that involve the orchestration and practical use of the simpler skills—like computation in school arithmetic—that we find so much easier to teach and to measure. Since we do not teach wholly, or primarily, for the ability to perform artificial school exercises, calls for the teaching and testing of higher-order cognitive skills are becoming more frequent. Today, theories of cognitive skills are emerging (at least within particular subject-matter domains) that sketch the specifics of the hierarchy of control behind our intuitive concept. In some cases, we are beginning to understand the course of skill acquisition as well. To the extent that such theories have been achieved, they provide a principled basis for our efforts to transmit higher order cognitive skills and for our efforts to determine whether we have succeeded.

In this chapter, I first characterize what we understand about higher-order skills in each of many domains, providing references to some of the major relevant theoretical statements. I then discuss an emerging consensus about the nature of the human cognitive system that implements and executes these skills. This section incorporates the discussion of notions about the changes that occur with learning and development together with implications for the kinds of skill which one might expect to see, or to deliberately train, in individuals of different ages. Next, discussion turns to the issue of the possible existence of very general cognitive skills in problem solving, knowledge acquisition, and in the control of one's own cognitive activities. How might such skills be understood within the theoretical framework that has been sketched? How might one teach them or test for them in individuals who do not have large amounts of domain-specific knowledge? This section concludes with a discussion of the possibility that we have at least two competing, and perhaps somewhat contradictory, notions of what is meant by higher-order cognitive skills, both deserving serious attention. Finally, the chapter concludes with a discussion of the ways in which these higher-order cognitive skills might be transmitted from one generation to the next.

DOMAINS OF COGNITIVE SKILL

Mathematics

Taking school mathematics as a first example, we all agree that arithmetic computations exemplify a lower-order cognitive skill. Notice that, even within computation, there are multiple levels of cognitive skill: There are skills in the coordination and orchestration of the simplest computations in order to carry out such algorithms as multicolumn subtraction or long division. But the important point is that the ability to do such computations—simple or complex—is of little use unless you known when to use them.

One of the less demanding uses for computational skills arises in the solution of word problems in school mathematics. Word problems require translation of a stereotyped situation into a set of appropriately selected computations. These translations are generally considered to involve higher levels of cognitive skill, and student performance on such problems is frequently disappointing (NAEP, 1983). In recent years, there have been substantial advances in our understanding of the processes involved in solving word problems (Chipman, 1988; Hinsley, Hayes, & Simon, 1977; Marshall, 1984; Mayer, 1981; Riley, Greeno, & Heller, 1983; Silver, 1985; and—dealing with more advanced mathematics—Schoenfeld, 1985). Yet even the ability to do word problems does not really measure a student's ability to use mathematics in the solution of life problems. Word problems present a variety of simplified and stereotypic sets of quantitative relations. To use mathematics effectively, one must be able to recognize situations in the world as being appropriate to mathematical treatment and be able to implement that treatment—abstracting out the quantitative relations from a myriad of irrelevant details.

Logic

Logic, like mathematics, is a formal computational system that many believe can be useful in ordering thought and solving problems. Although it is not usually taught as an elementary school subject, it, too, has relatively low-level symbol manipulation skills within the formal system. As with mathematics, the ability of people to use formal logic in solving even artificially set problems seems to be limited. Johnson-Laird (1983, 1985; Johnson-Laird & Bara, 1985) provides useful discussions both of shortcomings in logical thinking and of the processes that people actually seem to use when they are successful in dealing with logical problems. Many researchers doing developmental psychology in the Piagetian tradition have done research on propositional reasoning: for example, Demetriou and Efklides (1985) and Falmagne (1980). This work shows the continued development of logical reasoning ability in adolescence. The *Philosophy for Children* program (Lipman, 1985) has an unusual pedagogical approach, using specially written novels, which attempts to help students recognize the susceptibility of complex situations to logical treatment.

Computer Programming

Computer programming is the new formalism of our day, for which many have hopes of general mind-strengthening power (Papert, 1980). It, too, has a lower layer of skill which is most easily defined and assessed: the commands and syntax of a particular language. Because there is no one, agreed-upon computer language, the shortcomings of defining programming skill at this level have been particularly evident. In research projects aimed at understanding the nature of programming knowledge and its effects on thinking, efforts have been made to assess some higher levels of programming skill. For example, Linn (1985) suggested a hierarchy in which program design skills involving knowledge of stereotypic patterns of code used to perform commonly encountered tasks, and procedural skills for combining those templates, lie above the level of language features. Similarly, she suggested that more generalized problem-solving skills could be involved, that would apply across a variety of programming languages or other formal systems. Jeffries, Turner, Polson, and Atwood (1981), Kurland, Mawbry, and Cahir (1984), and Ehrlich and Soloway (1983), among others, have much to say about the structure of programming skills. More recently, Soloway, Spohrer, and Littman (1988) have proposed an approach to the teaching of programming in which students are required to produce programming plans, indeed several alternate programming plans, and are graded upon these as well as upon the success of the program itself.

Physics

The case of such scientific subjects as physics is comparable to mathematics. Textbook problems in mechanics present a formidable challenge to most students, even though they demand only an appropriate selection of physical laws and mathematical computations to deal with cleanly defined and idealized situations. But there is still a higher level of skill: The student will not be much use as a productive engineer, for example, unless complex and confusing situations can be translated into the terms of the science. Genuine understanding of scientific ideas involves understanding how those ideas project onto all the complexity of real situations, seeing the *regularity* that is taught in science courses as a figure that stands out in the background of rich contextual detail. Many researchers have contributed to our understanding of the higher levels of skill that guide the use of elementary knowledge, for example recognition of deeper structures in problems, translation of problems into scientific representations, and so on. (Chi,

Feltovich, & Glaser, 1980; Reif, 1985, 1986; Larkin, 1985; McDermott, 1984; Larkin, McDermott, Simon, & Simon, 1980).

General Scientific Reasoning

The Piagetian research tradition has much to say about certain aspects of scientific reasoning, particularly tasks involving the combinatorics and control of variables in experiments. Of course, it is true that these are relatively high-level skills as compared to the computations, data-recording acts and other activities that may be organized into these larger strategies. Much information is available about developmental progressions in such tasks (e.g., Siegler, 1976, 1985; Demetriou & Efklides, 1985; Arlin, 1975; Commons, Richards, & Kuhn, 1982; Fischer, 1980; Neimark, 1981).

Krupa, Sellman, and Jacquette (1985) outline interesting consistencies in the types of scientific explanations that seem to be preferred early versus late in the development of understanding of particular domains, patterns that should influence the design of assessments. It may be that what is critical is experience with the domain rather than general developmental status. The effect of specific knowledge of, and experience with, the content of these scientific reasoning tasks is an important factor in determining performance (Linn, Pulos, & Gans, 1981).

Altogether, the scope of what has been investigated and described as high level skill in scientific reasoning seems to be rather limited. In an exciting new book, (Langley, Simon, Bradshaw, & Zytkow, 1986), a much more complete view of the processes of scientific reasoning and discovery is outlined, within which the control of variables is seen as one of many relatively low-level strategies. Precursors to this work can be found in Hanson (1958).

Reading

The general idea of a hierarchical orchestration of skills can be used to think about what constitutes higher-order skill in other domains. In reading, for example, the decoding of letters into the sounds of the language is at the bottom of the hierarchy. Interpreting the contribution of syntactic structures and cohesive devices is at a somewhat higher level. Beyond that, there are discourse structures: those of stories, of expository texts, of arguments. Discourse skills are needed in order to appreciate why particular information is in the text. In the last 10 years, there has been a vast expansion of our understanding of these higher levels of reading skill. Recent books provide an overview of that progress, (e.g., Just & Carpenter, 1987; Orasanu, 1986; Van Dijk & Kintsch, 1983).

If we consider the utility of reading in parallel with the utility of mathematics, it becomes evident that there is a still higher level of skill that involves directing and controlling reading effort in order to meet the practical needs of the reader. A reader's practical needs can vary enormously. One might be gathering and organizing information to meet the demands of a particular task. One might be studying in order to grasp a particular area of content. One might be reading to evaluate the motivations and purposes of the writer in order to determine whether his or her messages and conclusions should be accepted. Skilled readers would deploy their lower-level skills of reading very differently in order to meet each of these very different goals. Relatively little research has been done on this aspect of high level reading skill; Smith (1967) is one of the few studies to address such questions.

Writing

The structure of writing skills shows strong parallels to the structure of reading skills. Correct spelling, punctuation, and grammar are at the bottom of the hierarchy. Stereotypic paragraph structure (main idea and supporting elaborations), argument structure, story structure, and so on, provide the next higher level up. Perhaps the most compelling evidence of poor or impaired higher-level cognitive skills has been uncovered by the efforts to assess these higher levels of writing skills: Student compositions have been found to be poorly organized and to be deteriorating in the quality of their organization (NAEP, 1981).

The work of Hayes and Flower (Flower & Hayes, 1981, 1982; Hayes & Flower, 1983) has shown the importance of still higher levels of writing skill—the flexible adaptation of these tools to the purpose of the writing effort and to the characteristics of the receiving audience. That is to say, if one observes the unfolding of the cognitive skill of writing in an expert writer, one observes first of all the setting of goals concerning purpose and audience, goals that orchestrate and control the lower-level skills that are actually involved in the construction of text. Most educated adults lack this level of skill, and that is why their efforts to write—business reports, device instructions, computer program documentation, and so on—are frequently not functionally useful.

History

In factual content areas such as history, it is less obvious than for mathematics what uses we expect for the knowledge. For some, the goal seems to be just to know particular facts for their own sake, or perhaps for the sense of one's place in a national and international tradition. Or is factual mastery merely the first level in a hierarchy of historical skills? Is the higher goal to be able to recognize instances of major historical phenomena—revolution, deterioration of a society, movements of national liberation, social stratification, and so on? Is it to be able to use these organized constellations of particular historical facts in order to be able to reason by analogy about the contemporary scene? To the extent that one can characterize these more complex accomplishments that are hoped for, we should be undertaking instruction that might be expected to foster them, and we should attempt to measure something beyond the mere memorization of historical facts when we assess the effectiveness of instruction.

Factually Rich Scientific Fields, Like Biology

Again, the question to ask is what the projected use might be of the factual knowledge in these fields. The answers will help to sketch the hierarchy of control in which factual knowledge has its place. Today, educators in biology want to see students able to use their knowledge in the design and interpretation of experiments, as discussed above in the section on general scientific reasoning. More research needs to be done in order to show how knowledge is brought to bear in the performance of such tasks. Those goals—experimental design and interpretation—focus on the training of future scientists and on general understanding of the scientific enterprise. But don't we also hope that people will be able to use their knowledge to think about their diets, about first-aid measures, about air and water pollution, about (in the military, for example) chemical and biological warfare protection? If so, what higher level skills will be required to orchestrate those facts in the service of such problem-solving activity? Will the concepts of the science be recognized as relevant? Research on the development of biological understanding is only beginning (Carey, 1985), and it is important to rely on the resources available in the field to specify what the major organizing concepts might be (Mayr, 1982).

THE FUNCTIONAL ARCHITECTURE OF HUMAN THOUGHT

The skills we have been discussing are embodied in the functioning of human beings. For the past 20 years especially, and long before, understanding the fundamental characteristics of the human system that thinks, learns, and solves problems has been a major scientific goal. Whatever questions we want to ask about cognitive skills, it is important to appreciate the characteristics of the human system that is implementing those skills. As we shall see below, the characteristics of the system interact with the characteristics of the tasks to determine the characteristics of the performance we see. Of course, questions about the fundamental nature of the human cognitive system remain matters of active scientific controversy and inquiry.[2] Nevertheless, consensus is emerging around a general theoretical framework that shows considerable power and success in explaining human cognitive performance at a level of description that is appropriate to guide our design of instruction, skill assessment, and the design of artificial systems for human use. In computer science, the term *functional architecture* is used to refer to the basic set of resources—both hardware and software resources—that are available in a computer system to support the creation of programs. Discussions of analogous concepts applied to understanding the human mind can be found in Pylyshyn (1984), Newell (1970, 1973, 1982), Anderson (1983), and elsewhere. There is general agreement that the functional architecture of human cognition has the characteristics described below.

Condition–Action Pairs

It is believed that the internal language of thought, *mentalese*, out of which the skills that interest us are built, can be characterized in terms of condition–action pairs. The actions of thought—whether internal cognitive actions or the occasional external visible actions that we observe—are taken when the existing situation is perceived to

[2] The presentation of the production system approach in this paper does not represent a commitment to the ultimate truth of this theory. At the present time, however, this theoretical approach appears to provide an appropriate level of explanation that is useful and productive in thinking about a wide range of cognitive phenomena. It seems likely that alternative theoretical accounts may provide a more detailed view of the way in which cognition is actually implemented in the human biological system, but that the various theories will approximate each other at the level of explanation relevant here.

match specified, previously learned conditions for those actions. So, the use of a lower-order skill like a simple computation would be triggered by the recognition of an appropriate situation. Among the conditions that govern actions is a particularly important case: The goal that the person is attempting to attain at the particular instance.

Limited Working Memory

An important feature of the human cognitive system is that this activity takes place within an internal arena of rather limited informational capacity called *working memory*. Only information that is in working memory controls the cognitive acts that are performed, many of which change the information contents of working memory, thus creating the conditions that will trigger further actions. Working memory functions as a narrow window on much wider bodies of information that might be relevant. This includes information coming through the perceptual system, some of which might be externally recorded information—such as a writer's outline or a problem-solver's diagram—useful in the thinker's task. It also includes information that is stored in long-term memory, such as general knowledge acquired in the past or hierarchies of goals (plans) that will guide the current activity. We are all familiar with the impression of "holding in mind" a limited portion of what we know as we work actively on a problem. This information in long-term memory may be very well learned and solidly established. Or it may have been hastily and insecurely stored—as when we work out a set of goals and supporting subgoals to accomplish a particular task. In the latter case, the information may be vulnerable to loss.

Note that this description implies several ways in which individuals could be more or less skilled in managing their own cognitive activity: in the way in which they scan external information, thus affecting what is available to internal cognitive activity; in the way in which they design and use external supports; and in making efforts to maintain the storage of plans and goals in long-term memory.

Flexibility of Response

Great flexibility of response to the circumstances of a particular situation is a consequence of the system characteristics just described. At the same time, the system will give an impression of organization and direction in its action. This is true because both established goals and features of the immediate situation enter into the determination of momentary action. Because momentary circumstances differ, it will be rare that exactly the same observable behavior is produced twice. Even haphazard differences in the information that falls within the range of visual attention may make a difference in the performance we see. Thus, higher-order cognitive skills appear to have characteristics such as those Resnick (1987) described: a nonalgorithmic character, complexity, multiple possible outcomes, application of multiple criteria, and evidence of nuanced judgment.

Goal Hierarchies as a Source of Control

Among the conditions that determine momentary action, there is typically a goal, or goals, within the arena of working memory. The orchestration of low-level skills that was described in the first section of this chapter can be managed by a hierarchical structure of goals and subgoals and sub-subgoals. The currently active goal in working memory would move up and down a hierarchy of goals stored in long-term memory. That is why the introductory section of this chapter refers to a hierarchy of control, not to a rigidly determined skill hierarchy. A goal hierarchy might be available in long-term memory as a result of previous experience and the development of expertise. Or it might have been developed during a preliminary analysis of the problem. In both cases, it would be subject to modification as performance proceeds. And in both cases, therefore, performance is subject to disruption if the goal information is not adequately maintained in long-term memory.

Let us consider a specific example. A writer might have a high-level goal of persuading a particular type of person to take a particular action, say to persuade a Mid-Western housewife to fasten her seatbelt every time she gets into her car. As an experienced writer, the writer might have in long-term memory a set of goals that reflect a stereotypic persuasive argument structure: State the issue and the alternatives, state positive arguments for the writer's position, state some negative arguments, explain those away, conclude in favor of the writer's position, and so on. At a lower level, there might be a goal of generating an example of a positive argument that is tied to the specific circumstances of the expected reader's life. Below that there would be a goal of writing a grammatical sentence that expresses the example in a vocabulary appropriate to the expected reader, perhaps eliminating terms like *torque* and *energy absorption* that might be in the research materials the writing is using. The currently active goals would shift up and down this hierarchy, moving in and out of working memory. At a particular moment, for example, we might catch the writer being concerned only with the crafting of a particular sentence.

If all of this is executed smoothly, we would see a higher-order cognitive skill in action. But it might be that the goals were never generated in the first place—in which case we would perhaps conclude that the person did not "have" the relevant skill, or perhaps did not exert the necessary effort to generate the goals. Or it might be that the goal information becomes lost from memory, or that the links to higher-level goals become lost. Then we might observe an opportunistic, associationistic flow of words governed only by low-level grammatical goals and the circumstantial content of the surrounding text. Writing, especially student writing, often looks like this.

Processes of Learning and Automatization

Processes of learning can modify both the conditions and the actions that are available within the cognitive system. With experience in successful execution, complex structures of action that were once executed in a slow, step-by-step fashion can become consolidated into one, large, rapidly executable action. This is called *automatization* (cf. Anderson, 1983; Schneider, 1985). Such consolidated actions greatly reduce the demands on working memory capacity and are much more resistant to disruption, at the price of being less flexible. Similarly, what were once fragmentary partial conditions for action can become consolidated into complex patterns that are rapidly and almost automatically recognized. This, too, reduces demands on working memory capacity.

Conditions as Conceptual Categories

The conditions that govern cognitive actions can be considered to be conceptual categories. That is, one is recognizing a situation in which prior experience suggests that particular actions might be effective. Consequently, the growth of expertise can be observed in the emergence of more sophisticated conceptual categories that govern action. Riley et al. (1983), Marshall (1984), and others have described general semantic schemata for word problems—categories of word problems—for which particular patterns of computational acts are appropriate. Similarly, Chi et al. (1980) have described expert and novice categorizations of physics problems that govern the direction of efforts toward solution. Notice that these conceptual categories are also the key to understanding generalization and transfer of learned skills: transfer will occur if the new situation is recognized as being an instance of the same category as the original situation, if both meet the conditions of application for the trained action. The research of recent years indicates that transfer of training to new situations is difficult, if not quite impossible, to find (Hayes & Simon, 1976; Bassok & Holyoke, 1990, for a recent summary).

Age-Related Changes in Memory Capacity

When we think about the development of cognitive skills in children and adolescents, it is important to recognize that there appear to be age-related changes in memory capacity (Huttenlocher & Burke, 1976; Chi, 1977; Pascual-Leone, 1970); that is, short-term and/or working memory capacity appears to increase very substantially from early childhood to adulthood. This changing capacity has considerable power in explaining which tasks children are and are not able to accomplish successfully (Pascual-Leone, 1970; Case, 1978, 1985). To some extent, tasks can be redesigned to reduce the demands on working memory, in which case one may obtain surprisingly mature performance from children or successfully teach them skills that had been considered beyond them (Case, 1985). One major explanation offered for this change in apparent memory capacity with age is just that everything is still being automated and chunked into larger informational units via the gradual learning processes mentioned above. For a child, a single syllable, for example, might take up as much memory capacity as an entire word or a stereotyped phrase does for an adult. Chi (1985) has shown that children are capable of much more mature than expected performance when the content of the task is highly familiar to them. Thus, the apparent change in memory capacity with age might simply reflect the fact that, on the average, children are much less familiar with the contents of any and all tasks that we might present to them.

These phenomena of apparently changing memory capacity suggest limits on children's potential for displaying higher-order cognitive skills. For example, a child might not be able to maintain in mind the multiple conditions involved in expressing a higher cognitive goal—such as adapting vocabulary to a reader while also supporting a point and satisfying grammatical constraints. There might just be an overflow of information. Or a child might be much more likely to lose information about links to higher-level goals. If one is attempting to teach or test for skill in the management and control of one's own cognitive activity, for example, tasks should be designed very carefully so that memory capacity isn't overwhelmed by low-level information necessary for performance of the task. Of course, keeping track of and managing what one is doing does place an extra burden on that capacity. Similarly, it is more likely that high-level skills will be displayed or can be taught if the content of the task is highly familiar; then, remembering task content information will not make excessive demands on working memory capacity, because the information will be in well-consolidated structures in long-term memory. Thus, Linn et al. (1980) have found that the display of Piagetian formal operational thinking depends upon the familiarity of task content. Voss, Greene, Post, and Penner (1983) have found similar effects on the quality of reasoning shown by adult political scientists. It is possible for skills and knowledge to be highly automated and consolidated only in restricted areas. For example, one could know a great deal about the Civil War and think in very sophisticated ways about the Civil War without knowing anything about other wars. One could be very automated and sophisticated in using whole-number arithmetic without being able to deal with decimals.

Implications of Automatization and Memory Limitations

The fact that automaticity resulting from extensive practice aids in overcoming memory limitations and thereby makes the performance of more complex tasks possible (Regian & Schneider, 1990) poses significant educational dilemmas. It is very natural for schools to focus on automating low-level skills—both for these reasons and because they provide simple, clear-cut goals to pursue. However, we must fight against the tendency to want to get absolutely everything nailed down at the low level before attending to anything at the high level. Notice that the condition–action theory suggests that it might be risky to practice low-level skills extensively without the higher level context of the conditions in which they are expected to be used. In addition, there is reason to believe that maintaining the relationship of low-level skills to important high-level goals may be necessary in order to provide motivation for learning (Heath, 1981). The scope of human knowledge is so broad that we cannot ever expect that everyone will have everything well mastered at the low level. Instead, we can hope that mastery of high-level cognitive skills within restricted domains will teach students what they should strive for as they master new areas throughout life. Sometimes, presumed priorities between low-level "basics" and higher-level skill may not be valid. One can imagine, for example, because there are many who illustrate it, that high-level proficiency in writing could be combined with poor mastery of spelling and violations of standard English grammar. To give another example, it is common to assert that high-level reading comprehension skills cannot occur until low-level decoding skills are mastered (Perfetti & Lesgold, 1979). This may be true in one's first language, but the experience is quite different when one starts to read a new language. Then, high-level comprehension skills remain relevant and can compensate for deficiencies in low-level skills. Therefore, the strategy of looking first for automaticity in low-level skills should be pursued with caution.

GENERAL COGNITIVE SKILLS OF PROBLEM SOLVING, KNOWLEDGE ACQUISITION, AND SELF-MANAGEMENT

One of the ongoing areas of controversy in both education and cognitive theory is whether higher level cognitive skills are necessarily domain specific, or whether there are very general, high-level strategies for the control of cognitive effort that can be applied across a wide range of human endeavors. The idea that we might be able to educate people in general problem solving or learning skills that would better equip them for whatever situations they confront in the future is highly attractive. Numerous educational programs exist with such goals (Segal, Chipman, & Glaser, 1985; Nickerson, Perkins, & Smith, 1985).

Two different conceptions of cognitive skill compete in this controversy. One conception might be characterized as the "cog wheel" or "Meccano set" notion of a skill: the idea that one can pick up a skill developed in one domain and reassemble it as a part in forming a skill relevant to another domain. The other conception might be called the "theoretical" theory: the notion that, although we see aspects of skills in various domains that look very much alike, to which we can give the same theoretical labels, each requires prolonged development grounded in the specifics of its domain—that there is no significant transfer of skills from one domain to another.

Within the theoretical framework of the condition–action rules just described, some resolution of this controversy can be sought. The generality of the skill arises from the generality of the concepts in its conditions. Just as the theorist can recognize a new instance of the "same idea," so might the thinker, who is, after all, a creature of the same species. The well-known patterns of concept formation should apply to this situation. It is quite possible that a potentially general skill might be learned with all sorts of peculiar and specific conditions attached to it. Problem-solving experience in several domains is much more likely, it would seem, to promote the formation of general problem-solving skills. On the other hand, it seems possible that very general concepts could be readily selected as the appropriate conditions for such skills, especially if those concepts were previously established or very "natural," such as "all thinking activities" or "all problem-solving activities."

Candidate General Cognitive Skills

If there are very general higher-order cognitive skills, what might they be?

Problem Solving

In problem solving, the so-called weak methods of problem solving (Newell, 1980; Langley, 1985) come immediately to mind: generate-and-test, means–end analysis, setting goals, making comparisons to the goal, and so on. It is significant that one sees these weak methods of problem solving most when (intelligent) novices confront a new domain (Simon & Simon, 1978). In fact, it could be that these general strategies are so general that they are part of the fundamental architecture of human thinking and might not be learned at all. Such a suggestion appears in Anderson (1987), who also suggests that these weak and general strategies have an essential, central role in the acquisition of domain-specific skills. The question of the possible innateness of these strategies has not, I think, been adequately investigated. Attempting to draw analogies to previously experienced cases also seems to be a very natural and almost universally applied strategy. Ongoing research in both programming (Pirolli & Anderson, 1985) and physics problem solving (Chi, Bassok, Lewis, Reimann, & Glaser, 1989) is showing the great extent to which students rely on example problems in texts as an analogical basis for attempting the solution of new problems. Very specific formal models of analogical reasoning have now been proposed (Gentner, 1983; Lewis, 1988). The formation of more distant problem-solving analogies has also been simulated in an artificial intelligence program (Larkin, Reif, Carbonell, & Gugliotta, 1988). Instructional programs that attempt to teach general problem-solving skills usually concentrate on other aspects—redefining goals and resources, thinking more creatively about what aspects of one's knowledge might be relevant to the problem, and so on. That is, they concentrate on setting up better working conditions for the general weak methods.

Knowledge Acquisition

Another interesting category of general skill—general by nature and definition—is the skill of being a good student, facile in acquiring knowledge about new subjects. Today, this skill gains new interest, because it is also the skill of the knowledge engineer. This skill, or these skills, have not been well defined or researched. Mostly, attention has been given to learning strategies that can be used to store information effectively in memory (cf. O'Neil, 1978; Segal, et al., 1985; Chipman, Segal, & Glaser, 1985). More recently, the seeking of *explanations*, sense-making cognitive activity, is coming to be seen as a key feature of human cognition (Miller, 1990). But there are also skills involved in determining what information ought to be stored, knowing such things as:

- Word meanings may be different from what you expect, and you should be sensitive to what context tells you about their meanings.
- There are likely to be important distinctions to be picked up, discriminations you may not be making.
- There is some set of concepts used to organize the subject matter—you should look for clues to the identity of those key ideas.
- You should organize the information you get systematically, and use various forms of representation to do so.

All but the last are almost uninvestigated, but perhaps this is an area in which we should be setting explicit educational goals.

Self-Management and Control of Cognition

It is almost universally agreed that the management and control of more specific cognitive processes is central to high levels of cognitive performance. As was pointed out in the previous section of this chapter, there may be cognitive skill involved in maintaining one's goals active in memory so that they can guide and structure performance. This could be done by constructing external aids, like outlines or flowcharts, or it could be done by periodically shifting attention to the mental rehearsal of one's plans, or it could be done in other ways. But all might be called *skills*. Beyond these relatively momentary forms of control, there is the idea of being organized and systematic in one's actions (is the control of variables schema really very different, for example, from the systematic search of a field that appears as an item on an individual IQ test?), of keeping track of the information gathered. There is the habit of being reflective about what one has done, and whether or not it was effective, and why: the habit of self-examination. Although these self-management activities might demand more working memory capacity than some children have available, it seems obvious that many do not practice them who could, and who could undoubtedly be taught to do so. Therefore, we may want to acknowledge these skills as a goal of schooling and to assess them.

Developing General Cognitive Skills

The fact that large amounts of specific knowledge are usually involved in interesting cognitive performances poses a challenge in any effort to develop general cognitive skills. The alternative of abstract or puzzle problems with low requirements for specific knowledge is worth considering. Such problems may also have the advantage of reducing memory loads to a minimum. These days it is somewhat fashionable to scorn the artificial puzzle problems that have been given so much attention by psychologists. However, it is germane to note that they are important artifacts of our culture, not just idle inventions of psychologists—*Scientific American* books of puzzles and games sell quite well. We can presume that they exist because they make thinking processes particularly clear, particularly easy to reflect upon and manage. They offer a kind of relevation of what thinking is about that many people seem to find enjoyable.

It is interesting, too, that Feuerstein (1980) has chosen to use very simple abstract materials in his thinking-skills program. Is it perhaps because these materials bring out the thinking processes that he is trying to teach, and that he can teach labels for those processes in a particularly sharp, salient, and unconfused way? Of course, the hazard of such teaching is that the skills might be learned as applicable only to artificial, abstract materials. Linking these abstract materials, and the thinking processes they evoke, to parallels in the students' life experience through classroom discussion is another important, but less easily communicated, aspect of Feuerstein's program.

If puzzle problems are an informal educational opportunity that is important in developing cognitive skills, and that is not equitably available, perhaps this is a circumstance that should be corrected. Of the two alternatives—abstract materials and puzzle problems built of everyday content—the latter are probably preferable. Johnson-Laird's (1983, 1985) work on logic would seem to indicate that the manipulation of abstract materials may require special skills, while people find it easier to build and think with mental models of familiar content.

Two Interpretations of "Higher-Order Cognitive Skill"

The foregoing discussion tends to bring out the fact that there are two different interpretations that we can give to the notion of higher-order cognitive skill.[3] One interpretation focuses on the depth and complex-

[3] This conceptual issue was pointed out by Steve Reder during the discussion at the San Diego conference.

ity of the skill in action, as exemplified by the discussion of domain-specific skills in the first section of this chapter. Part of the secret of complex skill development is that lower-level aspects of the skill become automated and no longer make demands on working memory capacity, so that there is a sense in which the complex skills are made relatively more simple. Correspondingly, the other interpretation which we can give to the idea of higher-order cognitive skill focuses on the top-level control mechanism for cognitive activity, on what is happening in working memory and how that is determining action. Much of that control mechanism is undoubtedly fundamental, innate cognitive architecture—a constant regardless of learning and experience. There is a sense in which both novice and expert performance in domain-specific skills may appear equivalent in the complexity of the cognitive control that we observe in working memory. However, as we have just discussed, there may also be learned skills that apply to the control of one's own cognitive activity. These will certainly interact with domain-specific skills to determine the sophistication of behavior that we ultimately observe. Both interpretations of *higher-order cognitive skill* point to important educational goals, and both make sense within the same cognitive science theoretical framework: They simply point to different aspects of the human cognitive system.

INTERGENERATIONAL TRANSMISSION

Having discussed what we mean by the higher cognitive skills, we can turn to the discussion of the way in which these skills can be developed in children as they mature into the adults of the next generation. There are at least two major modes by which the higher cognitive skills may be transmitted from one generation to the next. One is the informal mode—via informal interaction in the family and, later, on the job or in other social situations. The other mode is through formal education, in schools, colleges, and universities. It is probably correct to view the transmission of higher cognitive skills as the primary purpose of colleges and universities. Interestingly, something akin to the informal mode of apprenticeship learning tends to reappear at the most advanced stages of university education. A third mode of transmission should probably be mentioned. Books are written (and today other forms of communication like documentary films are created) that have the purpose of transmitting cognitive skills: Polya's (1945) *How to Solve It* as one example of many in that genre, numerous books on the arts of various kinds of writing, books by famous scientists about what they have learned about doing science, and so on. Perhaps this is a particularly chancy mode of transmission, demanding much of the recipient, but there is no doubt that many individual lives have been as dramatically influenced by the reading of such books as by a parent's or teacher's intellectual influence.

Informal Transmission

The influence of family members and of others close to the family comes first in everyone's life and continues to be important for many years. The power of socioeconomic status, educational status of parents, and similar variables in predicting school achievement and other intellectual outcomes undoubtedly testifies to the importance of the informal mode of transmission. There are many reasons why this might be so.

One of the most significant insights to have emerged from the cognitive science research of recent years that developed the theoretical framework described above is the realization of the great importance of "conditions of application" as part of cognitive skills. This aspect of cognitive skill is systematically neglected in most formal education. In mathematics or science, permissible problem-solving "operations" will be taught, but typically little or nothing will be taught about when it is appropriate to apply those operations. The consequence can be a skill that is useful only in school exercises—and sometimes not even there—unaccompanied by the ability to apply those skills to the real-life problems for which schooling was hoped to prepare the student. Thus, we are beginning to hear about the significance of *situated cognition*.

In informal learning situations, it is almost certain that an appropriate context will be present. At the least, the skill will not be taught in a way totally divorced from any context of application. At best, the mentor or model may comment upon the reasons for selecting a course of action and, having the context readily available, may point out the key features of the situation, controlling the learner's attention to them. As Collins, Brown, and Newman (1986) have pointed out, a cognitive apprenticeship is a bit more problematic than a conventional apprenticeship, because most of cognition is covert. But this varies with the individual person and, almost undoubtedly, with the cultural group. Some people like to voice their thoughts as they think through problems and decisions. A child's parents, in particular, may discuss together many of the problems and decisions that face them. They may even consult each other for advice in dealing with problems at work or with other people. Or simply hold forth with a discussion of their thinking, in search of sympathy, agreement, and moral support.

These are simple and rather obvious observations, but they have rather important implications. Only the skills which parents themselves possess and regularly exercise in their lives will be potentially available for transmission through an informal cognitive apprenticeship. Furthermore, whether or not that potential is realized may depend upon the personality of the parent and the cultural norms about expression that the parent observes, upon the amount of time that parent and child spend together, and, quite possibly, upon whether or not another parent is present to serve as a stimulant to overt expression of thoughts.

Most parents, perhaps all parents, regard themselves as teachers of their children. What they choose to teach, and why, and how, obviously can be of great importance. A parent who is inclined to transmit what he or she knows about many things—about scientific understanding of nature, for example, or about the operation of devices around the home—has much in common with an individual tutor. And we know that individual tutoring is a highly effective mode of instruction, capable of making the difference between outstanding and mediocre achievement (Bloom, 1984). What the "tutoring" by parents may lack in systematic plans and concentrated effort, it likely gains in coming at opportune motivational moments and in meaningful, perceptually rich contexts. Again, however, parents cannot transmit what they do not know.

Feuerstein (Feuerstein, Jensen, Hoffman, & Rand, 1985), one of the leading exponents of compensatory cognitive skills training, has been very eloquent on the subject of the importance of the *mediated learning experience* that parents provide as a factor in successful cognitive development. Adults, he says, select, filter, and interpret the stimuli that impinge upon the child, helping to make sense of the world. As the adult directs the child's attention and provides labels and interpretations, temporal and spatial order is imposed on what could be confusing. The interpretations provided by the adult introduce the categories and values of the child's culture, labeling the child's actions as signs of growing competence and relating actions to goals that are not a matter of material perception. The mediation of the adult, he believes, introduces the child to the use of cognitive representations of experience that have content beyond that provided by direct sensory experience. This view of the role of the social surround in creating the internal cognitive structure of the child is very similar to the increasingly popular views of Vygotsky (1962), the Soviet psychologist who studied the cognitive functioning of Soviet peasants during the 1930s, gaining much insight into the influence of culture upon cognitive function.

Some of the more general cognitive skills, the strategies for going about the tasks of learning themselves, may be among those that are both most neglected in schooling and least likely to be in the possession of many students. A government survey (USDE, 1982) has shown that schools expect "study skills" to be taught at home; yet, the majority of parents have no idea what study skills are. School personnel themselves, and similar educated people, do have considerable understanding of study skills and are able to help their children meet these implicit demands of schooling. Home efforts to help with the demands of schooling can vary dramatically in quality and quantity (Chall & Snow, 1982; Varenne, Hamid-Buglione, McDermott, & Morrison, 1982).

Unfortunately, it is possible that the influence of disadvantaged parents may amount to more than an absence of the ability to transmit specific school-related skills. In *Ways with Words*, Shirley Brice Heath (1983) argues that working-class white parents in the community she studied imparted a very rigid notion of the proper use of language, a combination of flat, factual reporting of events and memorized formulaic sayings, often Biblical quotations. This style, she felt, became an obstacle to the more flexible and imaginative use of language that school often calls for. These children, she says, did quite well in the very first years of school, when a memorization strategy would suffice, but began to flounder later on when the curriculum moved towards what we would call the higher-order cognitive skills.

Ethnographic studies like Heath's are rather subjective and impressionistic, but it is uncertain whether we can or need to do much better in capturing the complexity, detail, and cumulative impact of parents' influence on children's thinking. Efforts have been made to do controlled experimental/observational studies of the way in which parents interact with young children, especially when they try to teach the child to do something. These confirm the existence of differences in style among different social groups (McGillicuddy-deLisi, 1982). Although the correlation between what is observed in these snippets of behavior and general cognitive characteristics of the children are not usually impressive, there are isolated results that suggest the possibility of large impact. For example, Richards and Siegler (1981) showed that nonspecific encouragement to figure out reasons for events could have moderately lasting effects upon the systematicity of young children's problem-solving behavior. The sort of event that this experiment represents, repeated many times over, might well have a substantial cumulative impact upon a child's cognitive activity and resulting learned competence. Ongoing studies that attempt to understand consistent patterns of ethnic and gender differences in school achievement (e.g., Campbell & Mandel, 1988) may yield greater insight into such attitudinal influences.

In short, there are many reasons why we might expect the cognitive status and the cognitive skills of the child to be much like those of the parents. And that seems to be what we observe when we find that SES is so strong a predictor of things cognitive. Perhaps it would be the wiser course simply to resign ourselves and accept the consequences of all those forces as inevitable. But we live in a time of wrenching social and economic change, in which the demand for higher-level cognitive skills is increasing while there is less and less need for the person skilled in learning and executing rote operations within the plans and goals set by others. We have been quite successful in mechanizing the rote operations of work. There is a need to alter the slow evolutionary pace of change that informal intergenerational transmission of cognitive skills can provide. For that, we have no choice, it seems, but to turn to professional educators and institutionalized forms of education—even if those must differ significantly from what we have known in the past.

Formal Transmission

In the U.S. today, schools are viewed with much ambivalence. Perhaps this is no more true than when we consider them as a possible medium for the transmission of higher order cognitive skills. The schools are a place where all children are brought into contact with the educated middle class, creating an opportunity for the transmission of skills that may not be present in their home environments. Maybe. On the one hand, we are concerned that the mismatch between some of the children and the middle-class expectations of the school can cause a failure of that contact, a missed opportunity. On the other hand, we may be skeptical of the extent to which the teachers themselves possess the higher-order cognitive skills that we would like to see transmitted. Gifted as many individual teachers may be, we know that the profession on the whole is not attracting a very able segment of the population.

Furthermore, we know that the school is a special environment with its own stylized patterns of interaction. In the many-to-one situation of the school, constrained by prescribed methods and curricular goals, the teacher may not provide the same cognitive opportunities that he or she could provide in a less formal one-on-one situation. Indeed, we know that the cognitive opportunities for learning higher-level skills are not available equally to all, even within the same classroom (McDermott, 1974). The increasing emphasis on detailed skill testing and accountability may be creating an increasingly less favorable environment for the development of higher-level cognitive skills. The sad epilogue to *Ways With Words* (Heath, 1983) reports that the innovative teaching described in the book has faded, a change that the teachers attribute to the accountability movement and its increasing regimentation of teachers (but it could be that the innovations were just too much work to sustain). Inevitably, the focus tends to be on the lower-level skills that we know how to test. As yet, our understanding of higher-order cognitive skills tend to be too poor to allow us to construct effective measures of them.

Both testing and instruction that is systematically designed to permit students to pass the tests tend to focus on those aspects of cognition that researchers have found relatively easy to bite off and chew, a collection of elementary skills that do not necessarily add up to a functioning whole. The cognition that has received scientific attention has been decried as a "dismembered" cognition (Miller, 1985); if that is a matter of scientific concern, it is still more disturbing when that incomplete view is taken too seriously as complete guidance for the design of instruction. Hence, we see the rising concern about contextless learning of fragmentary skills, the rising enthusiasm for *situated cognition*, extolling the virtues of informal learning.

It is difficult to remember that, only a few years ago, the *context-free* character of school learning seemed to be one of its distinguishing virtues (Bruner, Olver, & Greenfield, 1966). Schools were seen as places where talk about things not present went on, where there was talk of things that might be. In this way, schools were considered to foster the formation of systems of cognitive representation, the ability to consider possibilities, to plan independent of the commitment and hazards of actual action, and so on. The critical issue, no doubt, is whether or not the children actually understand what is being talked about, whether they can in fact imagine the situations being described, or whether school talk becomes meaningless verbiage.

Perhaps mathematics presents the dilemma most starkly. It is clear that school mathematics is meaningless symbol manipulation for many. Equally, it has been recognized that people are generally capable of quantitative reasoning in the practical life situations in which they find themselves (Lave, 1988). Yet any mathematically educated person will cringe at the thought that these practical strategies are enough. Mathematics is one of the proudest achievements of human cultures. Its symbol systems—and their manipulations—provide immense amplification of human cognitive powers, permitting us to work through implications and conclusions that would overwhelm our informal reasoning powers. That is what we hope to transmit to the students in our schools. But we are less than successful in teaching students how that symbol system maps onto reality so that it can become truly useful. Clearly, that is not a reason to abandon it.

This, and the other discontents, are beginning to inspire new teaching ideas with more promise of success in using the schools as a medium for the transmission of higher cognitive skills. Bransford (1988) is exploring the use of videodiscs to bring rich perceptual contexts into the classroom, to approximate the way the parent or tutor might point out the application of mathematics to real situations. Many are exploring the idea that the teacher might model the thought processes involved in problem solving or in planning a piece of writing, thus providing the opportunity for cognitive apprenticeship, violating the formal perfection that has often been the standard of performance for a teacher. After an earlier attempt at directly teaching the components of mathematical skill derived from a cognitive analysis, Schoenfeld (1987) is experimenting with the creation of a group problem-solving environment that provides something like a cognitive apprenticeship opportunity in which such component skills can be more naturally learned. Sternberg and Martin (1988) describe a style of inquiry teaching—and the difficulty of achieving it—that many consider vital in fostering the development of thinking skills.

In general, one might make the point that a scientific analysis of cognitive skill need not be taken too directly as the plan for the teaching of that skill. For reasons of motivation, to maintain the relationship of the component skills to the goals they serve and to the other relevant *conditions of application*, one may want to teach more holistically. Being mindful of the analysis may simply help to ensure that the opportunity to learn everything important about the skill is there. The hazards of insisting on the mastery and automation of low-level skills before children are ever introduced to the higher-level context that makes them meaningful are evident. One can have somebody or something (a computer) assume responsibility for the low-level aspects of a task while the child practices the higher-level aspects. In Heath's book, this option appears as children dictating stories to the teacher; in a computerized Algebraland (Foss, 1987), the student can direct the computer in carrying out low-level manipulations.

Often, much can be accomplished by explaining the whys and wherefores, the possible value of what is going on in school. Too often, these secrets are not shared with children. Heath (1983) tells of overcoming some of the dysfunctional rules of the game that children bring with them to school by giving them a more sophisticated view that different rules apply in different places. We have learned that learning strategies are much more likely to be used once taught if we help children to understand their efficacy (Paris, 1988). Perhaps they would even tolerate drills for automaticity if they knew where those component skills fit into a more useful whole and are helped to understand the importance of automaticity. There is reason to believe that our success in teaching the low-level skills will be limited if we do not also teach the high-level skills that give them meaning and purpose.

The abandonment of the schools for some imagined world of cognitive apprenticeship seems a romantic alternative, but it is not a feasible one. For a few, wonderful opportunities might be made available; for the many, opportunities that now exist would be lost. The challenge is to capture the virtues of both formal and informal learning in some form that is practical within the unavoidable constraints of school learning.

REFERENCES

Anderson, J. R. (1983). *The architecture of cognition.* Cambridge, MA: Harvard University Press.

Anderson, J. R. (1987). Skill acquisition: Compilation of weak-method problem solutions. *Psychological Review, 94*, 192-210.

Arlin, P. (1975). Cognitive development in adulthood; A fifth stage. *Developmental Psychology, 11*, 602-606.

Bassok, M., & Holyoke, K. J. (1990, April). *Transfer of solution procedures between quantitative domains.* Paper presented at the annual meeting of the American Educational Research Association, Boston.

Bloom, B. S. (1984). The 2 Sigma problem: The search for methods of group instruction as effective as one-to-one tutoring. *Educational Researcher, 13*, 4-16.

Bransford, J. (1988, April) *Uses of macro-contexts to facilitate mathematical thinking.* Paper presented at AERA, New Orleans.

Bruner, J. S., Olver, R. R., & Greenfield, P. M. (1966). *Studies in cognitive growth.* New York: Wiley.

Campbell, J. R., & Mandel, F. S. (1988 April). *How parents foster technical development in minority girls.* Paper presented at the annual meeting of the American Educational Research Association, New Orleans.

Carey, S. (1985). *Conceptual change in childhood.* Cambridge, MA: MIT Press.

Case, R. (1978). A developmental based theory and technology of instruction. *Review of Educational Research, 48*, 439-469.

Case, R. (1985). *A developmentally based approach to the problem of instructional design.* In S. F. Chipman, J. W. Segal, & R. Glaser (Eds.), *Thinking and learning skills, Vol. 2: Research and open questions.* Hillsdale, NJ: Erlbaum.

Chall, J., & Snow, C. (1982, December). *Families and literacy: The contribution of out-of-school experiences to children's acquisition of literacy* (Final report on NIE-G-80-0086). Cambridge, MA: Harvard Graduate School of Education.

Chi, M. T. H. (1977). Age differences in memory span. *Journal of Experimental Child Psychology, 23*, 266-281.

Chi, M. T. H. (1985). Interactive roles of knowledge and strategies in the development of organized sorting and recall. In S. F. Chipman, J. W. Segal, & R. Glaser (Eds.), *Thinking and learning skills, Vol. 2. Research and open questions*. Hillsdale, NJ: Erlbaum.

Chi, M. T. H., Bassok, M., Lewis, R., Reimann, P., & Glaser, R. (1989). Self-explanations: How students study and use examples in learning to solve problems. *Cognitive Science, 13*, 145-182.

Chi, M. T. H., Feltovich, P. J., & Glaser, R. (1980). Categorization and representation of physics problems by experts and novices. *Cognitive Science, 5*, 121-152.

Chipman, S. F. (1988, April). *Word problems: Where test bias creeps in*. Paper presented at AERA, New Orleans. (ERIC :TM012411).

Chipman, S. F., Segal, J. W. & Glaser, R. (Eds.). (1985). *Thinking and learning skills. Volume 2: Research and open questions*. Hillsdale, NJ: Erlbaum.

Collins, A., Brown, J. S. & Newman, S. E. (1986). *Cognitive apprenticeship: Teaching the craft of reading, writing, and mathematics* (BBN Tech. Rep. No. 6459). (To appear also in L. Resnick (Ed.), *Knowing, learning and instruction: Essays in honor of Robert Glaser*. Hillsdale, NJ: Erlbaum.)

Commons, M. L., Richards, F. A., & Kuhn, D. (1982). Systematic and metasystematic reasoning: A case for levels of reasoning beyond Piaget's stage of Formal Operations. *Child Development, 53*, 1058-1069.

Demetriou, A., & Efklides, A. (1985). Structure and sequence of formal and postformal thought: General patterns and individual differences. *Child Development, 56*, 1062-1091.

Ehrlich, K., & Soloway, E. (1983). An empirical investigation of the tacit plan knowledge in programming. In J. Thomas & M. L. Schneider (Eds.), *Human factors in computer systems*. Norwood, NJ: Ablex Publishing Corp.

Falmagne, R. J. (1980). The development of logical competence: A psycholinguistic perspective. In R. H. Kluwe & H. Spada (Eds.), *Developmental models of thinking*. New York: Academic Press.

Feuerstein, R. (1980). *Instrumental Enrichment: An intervention program for cognitive modifiability*. Baltimore: University Park Press.

Feuerstein, R., Jensen, M. R., Hoffman, M. B., & Rand, Y. (1985). Instrumental Enrichment: An intervention program for structural cognitive modifiability. In: J. W. Segal, S. F. Chipman & R. Glaser (Eds.), *Thinking and learning skills, Volume 1: Relating instruction to research* (pp. 43-82). Hillsdale, NJ: Erlbaum.

Fischer, K. W. (1980). A theory of cognitive development: The control and construction of hierarchies of skills. *Psychological Review, 87*, 447-531.

Flower, L., & Hayes, J. R. (1981). A cognitive process theory of writing. *College Composition and Communication, 32*, 365-387.

Flower, L., & Hayes, J. R. (1982). Plans that guide the composing process. In C. Frederiksen, M. Whiteman, & J. Dominic (Eds.), *Writing: The nature, development and teaching of written communication*. Hillsdale, NJ: Erlbaum.

Foss, C. L. (1987). *Learning from errors in ALGEBRALAND.* (Tech. Rep. No. IRL-87-0003). Palo Alto, CA: Institute for Research on Learning.

Gentner, D. (1983). Structure mapping: A theoretical framework for analogy. *Cognitive Science, 7,* 155-170.

Hanson, N. R. (1958). *Patterns of discovery.* Cambridge, UK: Cambridge University Press.

Hayes, J. R., & Flower, L. (1983). Uncovering cognitive processes in writing: An introduction to protocol analysis. In P. Mesenthal, L. Tamon, & S. Walmsley (Eds.), *Research in writing: Principles and methods.* New York: Longman.

Hayes, J. R., & Simon, H. A. (1976). The understanding process: Problem isomorphs. *Cognitive Psychology, 8,* 165-190.

Heath, S. B. (1981). Toward an ethnohistory of writing in American education. In M. F. Whiteman (Ed.), *Writing: The nature, development and teaching of written communication. Volume 1: Variation in writing: Functional and linguistic cultural differences.* Hillsdale, NJ: Erlbaum.

Heath, S. B. (1983). *Ways with Words.* Cambridge, UK, and New York: Cambridge University Press.

Hinsley, D. A., Hayes, J. R., & Simon, H. A. (1977). From words to equations: Meaning and representation in algebra word problems. In M. A. Just & P. A. Carpenter (Eds.), *Cognitive processes in comprehension.* Hillsdale, NJ: Erlbaum.

Huttenlocher, J., & Burke, D. (1976). Why does memory span increase with age? *Cognitive Psychology, 8,* 1-31.

Jeffries, R., Turner, A., Polson, P. G., & Atwood, M. L. (1981). The processes involved in designing software. In J. R. Anderson (Ed.), *Cognitive skills and their acquisition.* Hillsdale, NJ: Erlbaum.

Johnson-Laird, P. N. (1983). *Mental models.* Cambridge, MA: Harvard University Press.

Johnson-Laird, P. N. (1985). Logical thinking: Does it occur in daily life? Can it be taught? In S. F. Chipman, J. W. Segal, & R. Glaser (Eds.), *Thinking and learning skills. Vol. 2: Research and open questions.* Hillsdale, NJ: Erlbaum.

Johnson-Laird, P. N., & Bara, B. G. (1985). Syllogistic inference. *Cognition, 16,* 1-61.

Just, M. A., & Carpenter, P. (1987). *The psychology of reading and language comprehension.* Rockleigh, NJ: Allyn & Bacon.

Krupa, M. P., Selman, R. L., & Jaquette, D. S. (1985). The development of science explanations in children and adolescents: A structural approach. In S. F. Chipman, J. W. Segal, & R. Glaser (Eds.), *Thinking and learning skills. Vol. 2: Research and open questions.* Hillsdale, NJ: Erlbaum.

Kurland, D. M., Mawbry, R., & Cahir, N. (1984, April). *The development of programming expertise in adults and children.* Paper presented at the annual meeting of the American Educational Research Association, New Orleans.

Langley, P. (1985). Learning to search: From weak methods to domain-specific heuristics. *Cognitive Science, 9,* 217-260.

Langley, P., Simon, H. A., Bradshaw, J. M., & Zytkow (1986). *Scientific discovery: An account of the creative process.* Cambridge, MA: MIT Press.
Larkin, J. H. (1985). Understanding, problem representation, and skill in physics. In S. F. Chipman, J. W. Segal, & R. Glaser (Eds.), *Thinking and learning skills. Vol. 2: Research and open questions.* Hillsdale, NJ: Erlbaum.
Larkin, J. H., McDermott, J., Simon, D. P., & Simon, H. A. (1980). Expert and novice performance in solving physics problems. *Science 203,* 1335-1342.
Larkin, J., Reif, F., Carbonell, J. G., & Gugliotta, A. (1988). FERMI: A flexible expert reasoner with multi-domain inferencing. *Cognitive Science, 12,* 101-138.
Lave, J. (1988). *Cognition in practice: Mind, mathematics and culture in everyday life.* Cambridge, UK: Cambridge University Press.
Lewis, C. (1988). Why and how to learn why: Analysis-based generalization of procedures. *Cognitive Science, 12,* 211-256.
Linn, M. C. (1985). The cognitive consequences of programming instruction in classrooms. *Educational Researcher, 14*(5), 14-16, 25-29.
Linn, M. C., Pulos, S., & Gans, A. (1981). Correlates of formal reasoning: Content and problem effects. *Journal of Research in Science Teaching, 18,* 435-447.
Lipman, M. (1985). Thinking skills fostered by Philosophy for Children. In S. F. Chipman, J. W. Segal, & R. Glaser (Eds.), *Thinking and learning skills. Vol. 2: Research and open questions.* Hillsdale, NJ: Erlbaum.
Marshall, S. P. (1984). Sex differences in children's mathematics achievement: Solving computations and story problems. *Journal of Educational Psychology, 76,* 325-336.
Mayer, R. E. (1981). Frequency norms and structural analysis of algebra story problems into families, categories, and templates. *Instructional Science, 10,* 135-175.
Mayr, E. (1982). *The growth of biological thought.* Cambridge, MA: Harvard University Press.
McDermott, L. C. (1984). Research on conceptual understanding in mechanics. *Physics Today, 37,* 24-32.
McDermott, R. P. (1974). Achieving school failure: An anthropological approach to illiteracy and social stratification. In G. Spindler (Ed.), *Education and the cultural process.* New York: Holt, Rinehart, and Winston.
McGillicuddy-deLisi, A. V. (1982). The relationship between parental beliefs and children's cognitive level. In I. E. Sigel (Ed.), *Parental belief systems* (pp. 7-24). Hillsdale, NJ: Erlbaum.
Miller, G. A. (1985). Dismembering cognition. In S. H. Hulse & B. F. Green, Jr. (Eds.), *One hundred years of psychological research in America* (pp. 277-298). Baltimore: Johns Hopkins University Press.
Miller, G. A. (1990). On explanation. In D. N. Robinson & L. P. Mos (Eds.), *Annals of theoretical psychology (Vol. 6).* New York: Plenum Press.
National Assessment of Educational Progress (NAEP) (1981). *Reading, thinking, and writing: Results from the 1979-1980 National Assessment of Reading and Literature* (Rep. No. 11-L-01). Denver, CO: Education Commission of the United States.

National Assessment of Educational Progress (NAEP). (1983). *The third national mathematics assessment: Results, trends, and issues* (Rep. No. 13-MA-01). Denver, CO.: Education Commission of the United States.

Neimark, E. (1981). Explanation for the apparent nonuniversal incidence of formal operations. In I. E. Sigel, D. M. Brodzinsky, & R. M. Golinkoff (Eds.), *New directions in Piagetian theory and practice*. Hillsdale, NJ: Erlbaum.

Newell, A. (1970). Remarks on the relationship between artificial intelligence and cognitive psychology. In R. Banerji & M. D. Mesarovic (Eds.), *Theoretical approaches to non-numerical problem solving*. New York: Springer-Verlag.

Newell, A. (1973). Production systems: Models of control structures. In W. Chase (Ed.), *Visual information processing*. New York: Academic Press.

Newell, A. (1980). One final word. In D. T. Tuma & F. Reif (Eds.), *Problem solving and education*. Hillsdale, NJ: Erlbaum.

Newell, A. (1982). The knowledge level. *Artificial Intelligence, 18*(1), 87-127.

Nickerson, R. S., Perkins, D. N., & Smith, E. E. (1985). *The teaching of thinking*. Hillsdale, NJ: Erlbaum.

O'Neil, H. F. (Ed.). (1978). *Learning strategies*. New York: Academic Press.

Orasanu, J. (Ed.). (1986). *Reading comprehension: From research to practice*. Hillsdale, NJ: Erlbaum.

Papert, S. (1980). *Mindstorms: Children, computers, and powerful ideas*. New York: Basic Books.

Paris, S. (1988, April). *Fusing skill and will: The integration of cognitive and motivational psychology*. Paper presented at AERA, New Orleans.

Pascual-Leone, J. (1970). A mathematical model for the transition rule in Piaget's developmental stages. *Acta Psychologica, 63*, 301-345.

Perfetti, C. A., & Lesgold, A. M. (1979). Coding and comprehension in skilled reading and implications for reading instruction. In L. B. Resnick & P. A. Weaver (Eds.), *Theory and practice of early reading* (Vol. 1). Hillsdale, NJ: Erlbaum.

Pirolli, P. L., & Anderson, J. R. (1985). The role of learning from examples in the acquisition of recursive programming skill. *Canadian Journal of Psychology, 39*, 240-272.

Polya, G. (1945). *How to solve it*. Princeton, NJ: Princeton University Press.

Pylyshyn, Z. W. (1984). *Computation and cognition: Toward a foundation for cognitive science*. Cambridge, MA: Bradford Books, MIT Press.

Reif, F. (1985). Acquiring an effective understanding of scientific concepts. In L. H. T. West & A. L. Pines (Eds.), *Cognitive structure and conceptual change*. New York: Academic Press.

Reif, F. (1986). *Cognitive guidelines for instruction: Applications to the teaching of scientific concepts*. Berkeley, CA: Department of Physics and School of Education.

Regian, W., & Schneider, W. (1990). Assessment procedures for predicting and optimizing skill acquisition after extended practice. In N. Fredericksen,

Glaser, A. Lesgold, & M. Shafto (Eds.), *Diagnostic monitoring of knowledge and skill acquisition*. Hillsdale, NJ: Erlbaum.

Resnick, L. B. (1987). *Education and learning to think*. Washington, DC: National Academy Press.

Richards, D. D., & Siegler, R. S. (1981). Very young children's acquisition of systematic problem-solving strategies. *Child Development, 52*, 1318-1321.

Riley, M., Greeno, J., & Heller, J. (1983). Development of children's problem solving ability in arithmetic. In H. P. Ginsburg (Ed.), *The development of mathematical thinking*. New York: Academic Press.

Schneider, W. (1985). Training high performance skills: Fallacies and guidelines. *Human Factors, 27*, 285-300.

Schoenfeld, A. H. (1985). *Mathematical problem solving*. New York: Academic Press.

Schoenfeld, A. H. (Ed.). (1987). *Cognitive science and mathematics education*. Hillsdale, NJ: Erlbaum.

Segal, J. W., Chipman, S. F., & Glaser, R. (Eds.). (1985). *Thinking and learning skills. Volume I: Relating Instruction to Research*. Hillsdale, NJ: Erlbaum.

Siegler, R. S. (1976). Three aspects of cognitive development. *Cognitive Psychology, 8*, 481-520.

Siegler, R. S. (1985). Encoding and the development of problem solving. In S. F. Chipman, J. W. Segal, & R. Glaser (Eds.), *Thinking and learning skills. Vol. 2: Research and open questions*. Hillsdale, NJ: Erlbaum.

Simon, D. P., & Simon, H. A. (1978). Individual differences in solving physics problems. In R. Siegler (Ed.), *Children's thinking: What develops?* Hillsdale, NJ: Erlbaum.

Silver, E. A. (Ed.). (1985). *Teaching and learning mathematical problem solving: Multiple research perspectives*. Hillsdale, NJ: Erlbaum.

Smith, H. K. (1967). The responses of good and poor readers when asked to read for different purposes. *Reading Research Quarterly, 3*, 53-84.

Soloway, E., Spohrer, J., & Littman, D. (1988). E UNUM PLURIBUS: Generating alternative designs. In R. E. Mayer (Ed), *Teaching and learning computer programming: Multiple research perspectives*. Hillsdale, NJ: Erlbaum.

Sternberg, R. J., & Martin, M. (1988). When teaching thinking does not work, what goes wrong? *Teachers College Record, 89*, 555-577.

U.S. Department of Education, Division of Management Systems Development, Organizational Performance Service, Office of Management. (USDE). (1982, December). *Study skills instruction: A service delivery assessment*. Washington, DC: Author.

Van Dijk, T. A., & Kintsch, W. (1983). *Strategies of discourse comprehension*. New York: Academic Press.

Varenne, H., Hamid-Buglione, V., McDermott, R. P., & Morrison, A. (1982). *The acquisition of literacy for learning in working class families* (Final report on NIE-G-79-0046.) New York: Teachers College, Columbia University.

Voss, J. F., Greene, T. R., Post, T. A., & Penner, B. C. (1983). Problem solving in the social sciences. In G. H. Bower (Ed.), *The psychology of learning and motivation: Advances in research and theory (Vol. 17)*. New York: Academic Press.

Vygotsky, L. S. (1962). *Thought and language*. Cambridge, MA: MIT Press.

Chapter 10

On the Intergenerational Transfer of Higher-Order Skills

Raymond S. Nickerson
Bolt Beranek and Newman Inc.

Susan Chipman's thought-provoking chapter, "The Higher-Order Cognitive Skills: What They Are and How They Might Be Transmitted," touches on several themes that are critical to an understanding of what is, or should be, meant by higher-order cognitive skills and of how they might be transferred from one generation to another. In this chapter, I shall pick up on a few of those themes and share some thoughts that they prompt.

THE DISTINCTION BETWEEN HIGHER- AND LOWER-ORDER COGNITIVE SKILLS

The distinction between higher- and lower-order cognitive skills is by now familiar to anyone who reads much in the educational research literature. As Chipman points out, the distinction gives us little trouble in casual discourse, but it can be a challenge when we try to be explicit about what we mean by it. Her attempt to be explicit about it emphasizes the orchestration or control function of higher-order skills and their involvement in the application of knowledge and lower-order skills in situations outside the classroom.

Chipman sharpens this characterization by illustrating the functioning of orchestration and control skills in several domains. In mathematics, she suggests, lower-order skills suffice for algorithmic computation, whereas higher-order skills underly the ability to apply mathematics effectively to real-world problems. In logic, symbol manipulation represents skills of a lower-order than does the ability to subject complex real-life situations to logical treatment. Lower-order skills involved in computer programming include mastery of the commands and syntax of a programming language; higher-order ones include those required in planning and program design. In physics, application of constructs of the domain to complex real-world situations requires skills of a higher order than does the solving of well-defined textbook problems under idealized conditions. Reading and writing can be viewed as processes that involve skills of various levels, ranging from letter or word recognition or production at one extreme to goal-directed purposeful reading or composition at the other. In traditional subject domains such as history and biology the acquisition of knowledge that defines the field demands a lower level of skills than does the ability to see the relevance of that knowledge to life outside the classroom and to apply it effectively there.

The distinction between higher- and lower-order cognitive skills should be seen, Chipman suggests, in the light of our current understanding of the architecture of human thought, because human performance is determined jointly by the characteristics of the human cognitive system and those of the tasks we perform. While the nature of the human cognitive system is still a matter of scientific investigation, Chipman sees a consensus emerging on several points.

- Human performance can be understood in terms of condition–action pairs—specific actions are triggered when perceived situations match learned conditions for those actions.
- Performance is constrained by the limited capacity of working memory.
- Performance is characterized by flexibility, which is not the same as lack of direction.
- Behavior is determined by multilevel goals. Not losing sight of higher-order goals while working toward the realization of subsidiary ones is a problem. Control in the interest of pursuing a goal is a prominent aspect of higher-order cognition.
- Automatization of complex action structures and of the recognition of complex patterns of conditions can enhance performance (at the cost of a loss in flexibility) by reducing demands on working memory.
- The conditions of condition–action pairs can be considered conceptual categories that can become more sophisticated as a consequence of learning.

- The capacity of working memory appears to increase with age during childhood and adolescence and this may help account for the increasing ability to accomplish more cognitively demanding tasks. The apparent capacity increase may reflect, in part, the effects of automatization and the organization of information in memory in larger chunks, which are consequences of familiarity with specific tasks.

This characterization of the differences between higher- and lower-order cognitive skills and their relationship to the architecture of cognition provides several foci for discussion. What I would like to pursue is the basic question of the merits of the distinction between higher- and lower-order skills. I believe this distinction has been a useful one, but I believe too that it lends itself to excesses of various sorts, most of which can be seen as instances of grossly oversimplifying the nature of human cognition.

As Chipman notes, one educational implication that is sometimes drawn from the distinction between higher- and lower-order skills, coupled with recognition of the effectiveness of automaticity for overcoming memory limitations, is the idea that all lower-order skills should be automatized through extensive practice before attention is given to the development of higher-order ones. Chipman challenges this view and questions the notion that lower-order skills are always essential prerequisites to the development of higher-order skills; there are people, she notes, who demonstrate a high level of proficiency in writing in spite of a lack of mastery of spelling, and even grammar. Resnick (1987a,b) makes a complementary point, arguing that "meaning-imposing" or "meaning-constructing" processes that involve higher-order thinking are essential to the development of real competence in such basic skills as reading and elementary mathematics as well as in other subjects that comprise the traditional school curriculum.

I want to suggest here the possibility that the distinction between higher- and lower-order skills is, or can be, misleading, and that we might do well to consider using somewhat different terminology. While neither Chipman nor Resnick has made this point explicitly, I believe it is consistent with the spirit of their observations. When we speak of higher- and lower-order skills, there is a tendency to classify any skill (process, operation, function, ability) that we can name as being in one or the other of these categories. So comparing, classifying, ordering are put in a low-level box, while decision making, problem solving, and reasoning are placed in a high-level one. But the fact is, it is possible to find instances of any of these skills that range from trivially simple and easy to execute to very complex and demanding indeed. One is comparing, for example, when one notices that one's socks do not match; one is also comparing when one details the similarities and differences between the Magna Carta and the U.S. Bill of Rights. And one is engaging in decision making when one decides which flavor of ice cream to buy just as surely as when one decides whether to purchase a new home or fix up the old one. Any generic skill can have many more and less complex instantiations.

This is a problem for the view that cognitive abilities can be represented adequately as a hierarchy in which the skills at one level are components of those at the next higher level. When we attempt to set up such a hierarchy, we discover that, no matter what we put in a box at a particular level, we are likely to be able to find examples of the same type of entity that is appropriately placed at some point higher or lower in the scheme (Nickerson, 1990). Moreover, given two skills, A and B, it is often equally easy to think either of A as a component of B or of B as a component of A. All this being true, assuming it is, might we not do well to think less in terms of higher- and lower-order skills and more in terms of higher- and lower-level realizations of any given skill, or of higher- and lower-level examples of specific cognitive functions? In characterizing different levels of cognitive activity, should we not use language that acknowledges that the same functions—or at least functions that are called by the same name—can be observed at various levels of complexity, rather than language that encourages us to associate specific functions with specific levels? We might then be inclined to distinguish between higher- and lower-order thinking, not on the basis of what the particular skills or processes involved are called, but on the cognitive demands imposed by the situation in which they are executed.

Attempts to distinguish between higher- and lower-level categorization, between higher- and lower-level concept formation, between higher- and lower-level decision making, between higher- and lower-level problem solving, could yield some new perspectives vis-á-vis the distinction between higher- and lower-order cognitive activity. Of particular interest is the question of whether examples of higher-order functioning have certain characteristics in common, independently of the skills involved. There are many possibilities that come to mind. To function at a high level means to understand, not only what one is doing, but why one is doing it; it means being in control of one's own cognitive resources; it means being reflective about one's task and able to judge one's performance in a relatively objective way; it means being flexible and able to adapt to changes in task demands; it means being able to see the commonalities between the task at hand and others that are analogous in substantive ways. Functioning at a high-level means having a realistic sense of one's own limitations. It means functioning with a level of confidence that is commensurate with one's ability. It involves attitudes, beliefs, motivation, and style, as well as skills, knowledge, and mastery of specific cognitive operations. It means being committed to certain standards of good thinking and striving to be fair in the weighing of evidence and the evaluation of arguments.

This view is compatible with the idea that people do not have to be taught to classify, to form concepts, to make decisions, to engage in problem solving, because we all do these things spontaneously, which is not to say that we do not need help in learning how to execute these functions in a higher level fashion than is our custom. The need is not to teach people to form conceptual categories, but to make them aware, not only of the necessity for this ubiquitous process and of its role in providing us with a conceptual universe that is sufficiently simple for us to cope with, but also of the quasiarbitrariness of the conceptual boundaries we draw and of the difficulties we can create for ourselves when we take those boundaries too seriously. The challenge is not to teach people to make decisions, but to make them less impulsively. It is not necessary to teach people to make inferences, but most of us could use some help in becoming fairer and more effective in the ways in which we seek and use evidence.

GENERAL VERSUS DOMAIN-SPECIFIC SKILLS

Chipman considers the question of whether there are general, or domain-independent, cognitive skills and looks for examples of such skills that relate to problem solving, to knowledge acquisition or to self-management. Certain approaches to problem solving—goal setting, means–ends analysis, use of analogies—seem to be applicable across domains and sufficiently widely used, even in the absence of efforts to teach them, she suggests, to justify the question of whether they are innate to the architecture of human cognition. With respect to general knowledge-acquisition skills, Chipman notes the possibility of acquiring knowledge about how information is represented and organized that could facilitate further knowledge acquisition. Management and control of one's own cognitive processes, she suggests, are almost universally seen as central to high levels of cognitive performance; control in this context includes moment-to-moment maintenance of one's orientation to a goal as well as habitual systematicity and reflectiveness in one's approach to cognitive tasks.

I agree with Chipman's observations here, and in particular with her position that one can identify important aspects of cognitive performance that are not unique to specific domains. I would go a little further and argue that some of the problem-solving heuristics that have been discussed and promoted by Polya (1957), Newell and Simon (1972), Hayes (1981), Schoenfeld (1985), and others, which are probably not innate to the architecture of human cognition, can be taught and are applicable in a variety of domains. These include breaking a problem into parts, finding one or more ways to represent a problem diagrammatically, and working backwards. A personal favorite, which I find useful in many contexts, is testing extreme cases.

Discussions of the question of domain-independent versus domain-specific skills seldom make clear what is meant by *domain independence* or *domain specificity*. If each of these terms is given an extreme connotation, we have, on the one hand, skills that are applicable in any conceivable domain and, on the other, skills that are applicable in a single domain only. I suspect there are many skills, or aspects of thinking more generally, that fall in neither of these categories. These are skills that one would hesitate to claim are applicable in all conceivable domains, but would see as applicable in more than one. Another question that is seldom addressed explicitly is what constitutes a domain. It seems to be the case that domains are tacitly equated with conventional school subjects, but this does not clarify things a lot. Is mathematics one domain, for example, or several, including arithmetic, algebra, and geometry? Is geometry one domain, or several, say plane geometry, solid geometry, and projective geometry? Note that these types of questions are bothersome only when we attempt to settle the issue of domain independence or specificity in a categorical way. We need to think, not in terms of a dichotomy (domain independence versus domain specificity), but in terms of range of applicability, and to recognize explicitly the possibility that this range can vary more or less continuously from very narrow to very broad.

Many people who have written recently about the nature of expertise have emphasized the importance of specific knowledge as an essential condition for highly skilled performance in any particular domain, however domain is defined. If the exercise of higher-order skills in a domain depends on having extensive knowledge of that domain, it would seem to follow that the teaching of such skills within specific domains would have to wait until the necessary knowledge had been acquired. Chipman urges that the alternative of using abstract problems or puzzles not be overlooked, but notes the importance of linking such materials and the thinking processes they invoke to concrete real-life analogues.

I believe the point is well taken. Abstract problems and puzzles are sometimes dismissed as irrelevant to the kind of thinking required in the day-to-day world, which typically involves solving domain-specific problems. But assuming, for the sake of argument, that no amount of experience with abstract problems will suffice to prepare one adequately to deal with all domain-specific challenges, it does not follow that such experience is of no value at all. Moreover, the question of whether what is learned in dealing with abstract problems transfers spontaneously to problem solving in domain-specific situations is different from the question of whether abstract problems can be used effectively in a training program designed explicitly to foster transfer. Abstract problems can be useful, in my view, for illustrating specific problem-solving heuristics in ways that are not complicated by the need for domain-specific knowledge, and for making the point that such heuristics are indeed domain independent to some degree. But if transfer to specific domains is the goal, the use of domain-independent problems for instruction should be coupled with illustrations of how the heuristics of interest can be applied in the targeted domains.

Transfer is an issue whether or not abstract problems are used for instruction. The evidence indicates that what is learned in a specific domain context typically does not transfer spontaneously to other domain contexts, even when it is appropriate to them. An argument that is sometimes made in favor of teaching generally useful heuristics in domain-specific contexts is that, even if transfer does not occur, at least one will have learned how to use those heuristics in one practical domain, and that argument carries some weight. If, however, one believes that transfer can occur and one wishes to facilitate its occurrence, one probably has to teach explicitly for transfer whether one uses abstract or domain-specific problems to do so. It is possible also, of course, to use both. A balanced approach to teaching general problem-solving methods might use abstract problems for introductory purposes and then follow the introductory treatment of particular methods with instruction regarding how to apply them in various domains.

SKILL TRANSFER

Chipman distinguishes two major modes in which the transfer of higher-order skills from one generation to another may occur: informally, through family and social or job situations; and formally, through educational institutions. The informal learning of skills is more likely than is the formal learning of them to occur in situations in which they are typically applied outside the classroom; this is an advantage, because it helps ensure that, along with the skills, at least some of the conditions of application will be learned. On the down side, informal learning from parents is limited by the skill levels the parents have attained, because one can pass on, through an informal cognitive apprenticeship, only the skills one has. It can also be constrained by such factors as the parents' personalities and the amount of time they spend with children.

Chipman points out the desirability of efforts by parents to help their children acquire effective study skills, but worries that most parents do not have a clear idea of what study skills are. She reminds us too of the unfortunate fact that the absence of ability to transmit something of positive value is not the worst situation one can imagine regarding parental influence on children's learning. There is also the possibility of passing on ideas and attitudes that have significant adverse effects.

Parents who are inclined to transmit what they know about various aspects of their world have much in common with tutors, in Chipman's view. Indeed, one might argue that teaching by a parent has a greater chance of being naturally tutorial than does teaching by a teacher in a classroom. It typically involves a teacher-to-student ratio of one to one; it can be highly interactive; the parent is likely to have a reasonable idea about what the child already knows and can adapt instruction to this reality; the child—at least in many instances—is motivated to learn from a genuine sense of curiosity and not simply because it is the expected thing to do.

Transfer Is Not Enough

The observation that parents, at least under certain conditions, have much in common with tutors reminds us of the enormous opportunity that parents who are prepared to play this role have. Teachers seldom have the degree of one-on-one access to children during their school years that a parent does during the years before school. Playing the tutorial role well, however, requires that a tutor possess the skills that are to be taught. While the question of how to facilitate the transfer of cognitive skills from parents to children (or from one generation to the next) is an important one, it would be a mistake to limit our attention to it. We want to know not only how parents can effectively pass on what they have learned, but also—and perhaps more importantly—how they can help children acquire skills and knowledge that they themselves do not possess. This is a special concern as it relates to disadvantaged groups. While the goal of helping parents to acquire cognitive skills that they can pass on to their children is clearly a desirable one, we would not feel comfortable accepting the parents' level of attainment as a ceiling beyond which the children should not be expected to go. In other words, while the facilitation of intergenerational transfer of cognitive skills is one objective toward which it makes sense to work, it is not the only one. We would also like to find ways to help parents to help their children acquire knowledge and skills that they themselves do not have and may not ever get.

What is needed may depend, of course, on the extent to which specific parents want their children to go beyond them. The kind of assistance that would be useful for parents who are keenly interested in having their children's intellectual development exceed their own would be of little use to parents who lack this interest. The challenge in the latter case is of a different nature entirely. Moreover, lack of enthusiasm for children's intellectual development can take more than one form. Contrast, for example, parents who simply have no interest in things literate or intellectual, either for themselves or their children, with those who are highly motivated to instill in their children specific beliefs and attitudes, but view the development of critical thinking ability as a threat to those objectives.

Skills Are Not Enough Either

While the passing on of skills from one generation to another is unquestionably important to the development of higher order thinking, it also is not enough. I would argue that skills are not the most important bequest that one generation can make to another. Of greater importance, in my view, are the attitudes (inclinations, dispositions), values, and beliefs that will motivate the development of skills and the humane and productive use of them.

Being a good thinker means, in part, being inquisitive, contemplative, reflective. It means being inclined to attempt to look at controversial issues—really look at them—from perspectives that differ from one's own. It means valuing objectivity and impartiality in the weighing of evidence, and—what is neither natural nor easy—being willing to subject favored hypotheses to critical evaluation. It means being desirous of understanding one's own biases and limitations as a thinker and of doing something about them.

Unlike the case of skills, here transfer may be the only process by which one generation can effectively influence a succeeding one. Parents are unlikely to be able deliberately to foster among their children the development of attitudes, values, and beliefs that they themselves do not have. One would hardly expect them to want to do so, but assuming there are such cases, it is not clear how one could go about this in a credible way; how does a parent say to a child, "I believe X but I hope you grow up to believe not-X?"

I strongly suspect that parents are much more influential than teachers in shaping attitudes, and that the results of this influence are long lasting. Parents who can help their children develop a sense of inquisitiveness, a love of learning, a commitment to fairmindedness, a respect for their own intellectual potential, and the like have given them a legacy that is of far grater value than any particular collection of skills that they could have passed on. This is not to deny the importance of the skills, but simply to state an opinion about priorities. Given the right attitudes, one has a good chance of acquiring the necessary skills on one's own; the converse seems to me considerably less likely to be true.

PARENTAL ROLES

Parents play many roles vis-á-vis their children. I will mention three that seem to me particularly relevant to the question of how they can transfer to their children, or help them acquire, the skills or whatever else is needed for higher-order cognitive functioning.

Sources of Information

Probably the most obvious difference between parents and young children, cognitively, is the fact that parents have a wealth of knowledge about the world that children have not yet acquired. The role of parents as information sources is so obvious that we may find it easy to overlook its importance. As Chipman points out, at least one leading exponent of the teaching of thinking has made this role central to his theory of cognitive development. Children who are deprived of the learning experiences that are normally mediated by parents or other adults who function as caretakers in parents' absence are likely, according to Feuerstein and his colleagues, to be delayed developmentally and to be in need of compensatory cognitive skills training (Feuerstein, Rand, Hoffman, & Miller, 1980). There is also the good possibility that children of parents who are not effective as information sources will arrive at school without some of the domain-specific knowledge about the world that "school-ready" children are typically assumed to have.

Agents of Control and Reinforcement

Parents set the rules, whether explicitly or not, by which families operate. They control, to a very substantial degree, what their children hear and see during the first few years of life; whether intentionally or by default, they open some doors on the world and not others. They determine what TV programs children get to see, what books are read to or by them, whether they learn what a library is, and so on. Parents convey to children certain expectations regarding behavior and, by reacting positively or negatively in various ways, encourage or discourage different forms of it. Assuming that parental approval is highly desirable to most children, and, conversely, that disapproval (or indifference) is painful, this function must be a highly influential one. This role differs in an important way from some of the others in that it is somewhat less knowledge dependent. One cannot be an effective source of information that one does not have, nor can one be a good role model for behavior that is beyond one's own competence; one can, however, encourage and reward the acquisition of knowledge and the development of skills that one is not oneself in a position to teach.

It is at least conceivable that how parents characteristically deal with questions, especially the more difficult ones, can have a profound and lasting effect on attitudes children acquire regarding various aspects of inquiry. I would guess that, depending on the way they habitually respond to children's questions, parents can either stimulate or stifle curiosity, they can either encourage or inhibit the development of good information-finding skills, and they can either promote the notion that every question has a simple straightforward answer that can be obtained by asking the right authority or foster a recognition that some questions can be answered only tentatively if at all.

Models

We know that intentional teaching is an important activity in many homes, and that many parents consciously and explicitly attempt to convey specific knowledge to their children and to teach them specific skills. There undoubtedly are also efforts to instill particular attitudes and values. One suspects, however, that most of what children learn in the home is learned, not because of deliberate efforts to teach, but rather as a consequence of incidental observation of the ways in which parents normally conduct their day-to-day lives, and of the patterns of behavior, habits, attitudes, and values they daily display.

HELPING PARENTS IN THOSE ROLES

The practical question that arises at this point is, what can be done to help parents and other people who have children in their charge to play these roles in such a way as to foster the development of the ability to function cognitively at a high level? There are at least two aspects to this question, one motivational and one intellectual.

I believe that motivation is a key variable in this context and that unmotivated parents constitute a very difficult challenge. But there can be no question that many parents are highly motivated and put considerable energy into enhancing their children's cognitive development as best they know how. Many other parents might become motivated if only they knew what to do.

What can parents do to foster the development of higher-level thinking ability in their children? And how should the answer to this question, assuming one can be constructed, be communicated to them? Here are challenges to researchers and educators alike. The question of how the development of the thinking ability of children is influenced by parents and early experiences in the home has not received the attention from researchers that it deserves, so while there may be a general consensus that the influence is strong, empirically substantiated details are lacking. As Calhoun (1973) notes, the prevailing attitudes about child rearing and beliefs regarding how family life affects the intellectual development of children have varied considerably over time.

There is a need, in my view, for the articulation, for the benefit of parents, of what is known about how the development of cognitive abilities is influenced by early home life, and in as specific a way as possible, what parents can do in the interest of enhancing the development of higher-order abilities. Communicating scientifically derived information to parents in such a way that they will find it useful is not something researchers, as a group, have been very good at doing. Researchers, including educational researchers, tend to talk primarily to each other. There exist examples of efforts by the educational research community to address itself to parents, but they are not numerous, and there is little basis for judging how successful they have been. Unfortunately, there are many examples of pop psychology books, written, usually, not by researchers, that purport to bring the findings of scientific research to the general public that present those findings in misleadlingly oversimplified, if not grossly distorted, ways. There is a need for writers who understand educational research and who can present its findings and their practical implications in a straightforward, useful and accurate fashion.

REFERENCES

Calhoun, D. (1973). *The intelligence of a people.* Princeton, NJ: Princeton University Press.

Feurstein, R., Rand, Y., Hoffman, M., & Miller, R. (1980). *Instrumental enrichment*. Baltimore: University Park Press.

Hayes, J. R. (1981). *The complete problem solver*. Philadelphia: The Franklin Institute Press.

Newell, A., & Simon, H. A. (1972). *Human problem solving*. Englewood Cliffs, NJ: Prentice-Hall.

Nickerson, R. S. (1990). Dimensions of thinking: A critique. In B. F. Jones & L. Idol (Eds.)., *Dimensions of thinking and cognitive instruction: A review of research* (Vol. 1). Hillsdale, NJ: Erlbaum.

Polya, G. (1957). *How to solve it* (2nd ed.). New York: Doubleday.

Resnick, L. B. (1987a). Constructing knowledge in school. In L. S. Liben (Ed.), *Development and learning: Conflict or congruence?* Hillsdale, NJ: Erlbaum.

Resnick, L. B. (1987b). *Education and learning to think*. Washington, DC: National Academy Press.

Schoenfeld, A. (1985). *Mathematical problem solving*. New York: Academic Press.

Chapter 11

Intelligence*

Richard Wagner
Florida State University

INTELLIGENCE: PAST, PRESENT, AND FUTURE

It is difficult to think of a psychological concept that has had a greater influence on our lives than that of intelligence. As students, all of us were given some kind of IQ or aptitude tests, and for many of us, important decisions concerning our education (e.g., whether we belonged with the high or low group) were made in part on the basis of our performance on such tests.

The field of intelligence, like all fields, has periods of dormancy and periods of activity. After emerging from a prolonged period of dormancy, the field has undergone profound change in recent years and will in all probability continue to change dramatically for at least the next decade. Given the goal of knowledge synthesis, especially with an eye toward recommendations for policy, practice, and future research, it may be helpful to consider past and present conceptions of intelligence, and to consider where conceptions of intelligence appear to be headed in the near future. Accordingly, this article is divided into three main sections, one each to deal with conceptions of intelligence of the past, present, and future.

* Preparation of this chapter was supported by Contract MDA90385K0305 from the Army Research Institute.

THE PAST: THREE CONCEPTIONS[1]

From the beginning of this century through the 1970s, three conceptions of intelligence dominated the field: the psychometric, the Piagetian, and the information processing.

The Psychometric Conception of Intelligence

The psychometric conception of intelligence grew out of the work of Binet (Binet & Simon, 1905) and the later work of Terman (Terman & Merrill, 1960) and other American psychologists. Psychometricians took an individual differences approach in their attempt to understand the *structure* of intelligence. Their most powerful tool has been the statistical technique of factor analysis. Factor analysis reduces individual differences variance across a set of variables or test scores into a smaller set of underlying factors. The underlying factors are conceptualized by psychometricians as latent mental abilities. Factor analysis began with Galton's (1888) concept of latent factors that arose in his explorations of relations between intelligence and eminence of both individuals and of cultures. Pearson (1901) contributed the mathematics necessary for constructing models of latent factors.

Spearman (1904) began modern factor analysis with his methods for analyzing the correlations among various aptitudes into a general factor (g), and task-specific factors (s), one for each task or test, that were not of interest in their own right. For Spearman, intelligence was represented by a single factor, g. Thurstone (1938) later proposed that intelligence consisted of not one but seven factors or primary mental abilities—verbal comprehension, verbal fluency, number, spatial visualization, perceptual speed, memory, and reasoning. Much of the controversy thereafter concerned whether intelligence was represented by one or multiple factors. Today, the consensus is that intelligence is best conceptualized was a hierarchy of mental abilities (e.g., Snow, 1978; Vernon, 1971). For example, Vernon's hierarchy consists of a general factor (g) at the highest level, two major group factors (verbal-educational ability and spatial-mathematical ability) at the next level, followed by minor group factors at the next level and specific factors at the bottom.

[1] The interested reader who desires an expanded treatment of this topic is encouraged to consult Wagner and Sternberg (1984).

The Piagetian Conception of Intelligence

The origin of the Piagetian conception of intelligence was Piaget's interest, while working in Binet's lab, in children's wrong answers to Binet's test items. Piaget, who was a biologist by training, was interested, not only in the structure of intelligence, but in its *function* as well. Piaget saw the function of intelligence as being identical with that of other biological functions, namely, adaptation to one's environment. Concerning the structure of intelligence, Piaget identified four major stages of intellectual development: the sensorimotor stage of infancy, the preoperational stage of early childhood, the concrete operational stage of middle childhood, and the formal operational stage of adolescence and adulthood. Moving from one stage to the next involves reorganizing and extending the cognitive structures of the earlier stage.

The Information-Processing Conception of Intelligence

Information-processing conceptions of intelligence are concerned with the cognitive processes and the representations the processes operate on when individuals perform intellectual tasks. Just as the factor is the fundamental unit of analysis for psychometrists, the *elementary information process* is the fundamental unit of analysis for information processing psychologists (Newell & Simon, 1972). What makes an information process elementary is that it is not subdivided into smaller units by the theory in question. Thus, what is considered to be the elementary level can vary from theory to theory.

Sternberg (1980), for example, proposed three kinds of information processes that were relevant to intellectual functioning: performance components, metacomponents, and knowledge-acquisition components. Performance components are the basic processes that are used to carry out task performance. Examples of performance components used in solving verbal analogy problems, problems such as DOCTOR is to PATIENT as LAWYER is to (a) JUDGE, (b) CLIENT, include encoding the terms of the analogy, inferring the relation between the first and second terms, mapping the relation to the third term, application of the relation to the answer options, and justification of answer choice when none of the options are judged to be completely correct. Metacomponents are higher-order processes that monitor and control the execution of performance components. Examples of metacomponents that are used in solving verbal analogies include selecting performance components, selecting knowledge representations, selecting a strategy for combining components, deciding whether to maintain a strategy, selecting a speed–accuracy trade-off, and solution monitoring. Knowledge-acquisition components are processes that are used to acquire knowledge. Examples include selective encoding, or sifting the relevant information from the irrelevant; selective combination, or combining selectively encoded information so as to form a coherent picture; and selective comparison, or selectively relating newly acquired information to old information.

Evaluation of the Psychometric, Piagetian, and Information Processing Conceptions of Intelligence

The three conceptions of intelligence should not be viewed as contenders for a single crown, in that they address different, but complementary aspects of intelligence. If a set of problems is given to children of different developmental levels, variance in performance can be partitioned into three sources: subject variance (i.e., individual differences), stimulus variance (i.e., item differences), and variance between developmental levels (i.e., developmental differences). Psychometric factors reflect subject variance. The concern is with latent abilities that explain why some individuals do well and others do poorly. Information-processing components reflect stimulus variance. The concern is with information processes that explain why some items are easy and others are difficult. One needs to be concerned with both subject and stimulus variance, and thus employ both psychometric and information-processing methods, for the following reasons. Any aspects of task performance that vary only trivially across individuals, yet nevertheless may be crucial for task performance, would be missed in factor analysis but would figure prominently in information processing analyses. Conversely, aspects of task performance that vary only trivially across items, yet nevertheless are important sources of individual differences in task performance, would be missed by information processing analyses but would figure prominently in factor analyses. Thus, in practice, there is reason to factor analyze parameter estimates from information-processing models, and reason to use information processing methods to decompose tasks that are of interest because of their factor structure.

The sources of variance that the psychometric and information-processing approaches tend to neglect, that the Piagetian approach does not have, are developmental differences. The Piagetian approach provides detailed descriptions of intellectual performance at different developmental levels, and concepts that relate to mechanisms of development (e.g., equilibrium, assimilation, and accommodation). It is, however, possible to study developmental differences using the psychometric approach by exploring differences in the factor structure of tasks across levels of development. It also is possible to study developmental differences using the information-processing approach by exploring differences in parameter estimates across levels of development.

There are at least two general limitations of the three conceptions of intelligence. The first is that they appear to account for only a subset of the intellectual competencies that are required in real-world as opposed to laboratory tests. Neisser (1976) refers to the typical problems found on IQ tests and in school as academic problems. It is argued that the academic problems differ from real-world or practical problems in a number of important characteristics. For example, academic problems tend to (a) be formulated by others, (b) come with all information needed for problem solution, (c) be disembedded from an individual's ordinary experience, and be of little intrinsic interest, (d) be well defined, in that there is no question just what the problem is, and (e) have only one correct answer, and one method for obtaining the answer (Neisser, 1976; Wagner & Sternberg, 1984). In contrast, one must often formulate one's own practical problems, and rarely is it clear exactly what the problems are, or in some cases, that there is even a problem at all. One must usually search for additional information to solve practical problems, and there typically are multiple "correct" solutions, each with its own liabilities and assets, and multiple methods for obtaining each. None of the three approaches to the study of intelligence represented by the psychometric, information processing, and Piagetian conceptions have dealt seriously with tasks and problems that are relevant to the everyday world beyond the classroom.

The second general limitation concerns relations between the conceptions of intelligence and learning and education. I will consider the limitations of each conception in turn.

Limitations of the Psychometric Conception of Intelligence for Learning and Education. Although the psychometric approach to intelligence has been a part of schooling ever since Binet's development of a test to identify students who were not smart enough to profit from a public school education, and although learning ability has been, and continues to be, an aspect of some definitions of intelligence (see e.g., Brown & Campione, 1986; *Journal of Educational Psychology*, 1921; Sternberg & Berg, 1986), the psychometric conception has not been particularly helpful to educators, beyond providing a very rough index of general level of cognitive functioning. One basic reason for this state of affairs is that the tests do not provide the kind of specific information that is needed to train students in the processes and products of learning.

A second basic reason for the limited usefulness of the psychometric approach to intelligence for education is that IQ tests sample only a narrow slice of the range of competencies that are required for effective school learning, let alone learning in out-of-school contexts (Wagner & Kistner, in press; Zigler & Seitz, 1982). This observation might appear to be in conflict with the solidly established empirical relation between IQ scores and achievement test performance. However, it is important to remember that the IQ-achievement relation appears to be a two-way street, in that not only does IQ influence school learning, but school learning also plays a crucial role in the development of intelligence, especially of the verbal abilities that are the strongest correlates of school achievement. Thus, it is not unlikely that a broad set of competencies influence school learning, which in turn influences performance on IQ tests. This would result in a correlation between IQ and achievement even if IQ sampled only a small portion of the competencies that were responsible for individual differences in school learning.

Concerns such as these have resulted in rethinking the appropriateness of IQ alone as a yardstick of the effectiveness of childhood intervention programs such as Head Start (Zigler & Seitz, 1982). IQ, or more specifically, pretest to posttest gain in IQ scores, has been the most widely used outcome measure of childhood intervention programs. The two obvious advantages of using IQ as opposed to other outcome measures are that the IQ test is probably the most reliable and objective instrument ever constructed, and that IQ is such a good predictor of achievement test performance. Another practical advantage is that it was soon learned that almost any childhood intervention program, even one lasting as short as 6 weeks, could be counted on to raise IQ scores 10 points, forgetting for the moment that such gains soon washed out when the program was finished (Eisenberg & Connors, 1966; Zigler & Seitz, 1982).

Zigler and Seitz (1982) have pointed out a number of disadvantages of relying solely on IQ to evaluate the effectiveness of childhood intervention programs. First, there are a number of factors that determine successful functioning both within and beyond the classroom, and although there is a predictive relationship between IQ and scores on achievement tests (tests that are little different than the IQ tests themselves), a very high IQ score is not enough to insure success in school. Second, there is ample evidence that among samples of retarded persons, factors such as personality and social competence are more important than IQ in accounting for individual differences in life adjustment (Penrose, 1963; Sarason, 1953; Tizard, 1953; Windle, 1966). One should not conclude that IQ is unrelated to life adjustment from these studies, because samples of retarded persons obviously represent a restricted range in IQ. What they do show is that other competencies also matter.

Limitations of the Piagetian Conception of Intelligence for Learning and Education. Although the Piagetian conception of intelligence has been of interest to educators (see, e.g., Furth, 1970), specific educational implications have been very difficult to derive. In fairness to Piaget's theory, it was not intended to be a theory of learning but a theory of intellectual development. There are a number of characteristics of the theory that make it ill suited for deriving specific implications for learning and education. First, the time ranges of the developmental periods are too great to be useful in sequencing instruction or training. Second, the theory emphasizes maturation, that is, the unfolding of a preprogrammed set of skills over time, rather than learning. Third, the empirical support for major aspects of the theory has eroded over the years (see, e.g., Brainerd, 1978; Brown & Desforges, 1979; Siegel & Brainerd, 1978).

Limitations of the Information-Processing Conception of Intelligence for Learning and Education. The information-processing approach would seem to be the most fertile of the three for deriving implications for learning and education. It is possible to use the approach to identify components that appear to play a role in performance on specific educationally relevant tasks. However, there are a number of important limitations. First, with the important exception of metacomponents, which by their very nature are quite general, the performance components that are identified with the approach tend to be more task-specific than is ideal. Thus, a serious attempt to carry out information-processing analyses of even a sample of educationally relevant tasks is likely to produce an unwieldy list of performance components that are potential candidates for training. Second, there is a wide gap between theory and practice, in that it is easier to carry out information-processing analyses of tasks than it is to design effective training programs for improving level of performance.

Summary. I have reviewed the three major previous conceptions of intelligence, the psychometric, the Piagetian, and the information processing. The three conceptions are complementary in that they address three different aspects of performance. The psychometric conception of intelligence has sought to account for differences in performance across individuals (subject variance). The information-processing conception of intelligence has sought to account for differences in performance across items (stimulus variance). The Piagetian conception of intelligence has sought to account for developmental differences in performance.

There are two basic limitations of these conceptions. The first is that their support derives from an empirical base that is limited to a very narrow range of academic-type tasks. It is this limitation that is being addressed by the present research, which will be described in the next section of this article, research that looks beyond IQ-type tasks. The second limitation is that the conceptions have not dealt adequately with learning and education. It is this limitation that is likely to be dealt with by research in the near future, a topic that will be addressed in the final section of this article.

THE PRESENT: BEYOND IQ

For most of this century, the study of intelligence has meant the study of intelligence tests. One unfortunate consequence is that the focus has been on measurement issues rather than theoretical ones. I recently reviewed the intelligence chapter in one of the most successful introductory psychology texts (Atkinson, Atkinson, Smith, & Hilgard, 1987). Unlike any other chapter in the text, including other chapters on subjects for which measurement is an important topic (e.g., personality), the first three-fourths of the chapter were devoted to issues concerning the measurement of intelligence; only at the end was cursory attention given to the most important question, namely, what it is that one should try to measure. Measurement issues are important, but so is a theoretical basis for the construct being measured. A second unfortunate consequence is that, because the early intelligence tests were constructed to predict success in formal schooling, subsequent research has focused on the academic domain to the neglect of intelligence as it operates in the everyday world.

Although there are several areas of active research in the field of intelligence at present, the one that appears to me to be most relevant to the present topic is that of practical intelligence. I will provide a brief survey of some recent developments and current issues in this area, and then provide some results of a study we recently carried out of induction.

Working Definitions of Practical Intelligence

Outright attempts to define intelligence have not been very useful, so it is unlikely that any definition of practical intelligence will be of anything other than heuristic value for the near term. However, examining working definitions can help to identify the assumptions associated with different approaches to the study of practical intelligence. Three working definitions of *practical intelligence* can be identified.

The first is to define practical intelligence by exclusion, that is, defining what it is not. Frederiksen (1986), for example, considers practical intelligence to be reflected in one's cognitive response to almost anything that happens outside the formal classroom setting. A related yet distinct approach to defining practical intelligence considers the characteristics of practical tasks that make them distinct from academic-type tasks. Thus, a practically intelligent individual is one who excels in solving problems that (a) are ill defined, (b) are incomplete, (c) have multiple solutions, and (d) have multiple means of reaching each solution.

A second working definition of practical intelligence is in terms of practical know-how. For example, Berry and Irvine (1986) have done a cross-cultural analysis of practical know-how, finding that it is widespread among peoples of the world, even peoples who have been regarded as savages by Western culture. Berry and Irvine conclude that examples of practical know-how meet the most important standard for defining intelligent behavior: behavior that permits the continuing adaptation and survival of the group. Wagner and Sternberg (1985; Wagner, 1987) consider facile acquisition of tacit knowledge to be the hallmark of practical intelligence. Tacit knowledge refers to practical know-how that usually is not taught directly and rarely is spoken about. In a number of studies, they have found tacit knowledge to predict career performance in professions such as academic psychology, business management, and sales, and that IQ is not an especially good predictor of facile acquisition of tacit knowledge. Although much of this work has involved tasks other than school learning, it is likely that facile acquisition of tacit knowledge is a determinant of success in formal schooling as well (Wagner & Kistner, in press).

A third working definition of practical intelligence centers on social competence (Ford & Miura, 1983; Zigler & Seitz, 1982). For example, Mercer, Gomez-Palacio, and Padilla (1986) consider practical intelligence to be meeting the normative expectations that others have for persons playing their particular role in their particular group. In a large-scale cross-cultural study involving nearly 2,000 children, Mercer et al. (1986) evaluated the construct validity of a measure of social-role performance of children. In addition to a number of subtests that assessed a child's ability to fulfill their roles in various contexts (e.g., family, community, school), they also gave the subtests of the Wechsler Intelligence Scale for Children-Revised. They found strong intercorrelations among the social-role subtests and among the WISC-R subtests, but that the correlations between the social-role measure and the WISC-R were nil. This suggests that the kinds of intellectual competencies required for formal schooling are different from those required in our interactions with others.

Proposed Relations Between Academic and Practical Intelligence

One cannot deny the existence and importance of the academic abilities that are measured by the traditional IQ tests. It is therefore important for practical intelligence theorists to consider how practical and academic intelligence are related.

Gardner's (1983) multiple intelligences framework identifies seven kinds of intelligence to date, with the possibility that more will subsequently be identified: musical, bodily kinesthetic, logical-mathematical, linguistic, spatial, interpersonal, and intrapersonal. Of the seven, three (logical-mathematical, linguistic, and spatial) would seem to be traditional academic abilities, two (interpersonal and intrapersonal) would seem to be practical abilities, and two (musical and bodily kinesthetic) would seem to be unique to the framework. Thus, a given kind of intelligence can be academic in nature, practical in nature, or neither. One problem for this approach is deciding whether there are seven or seventy kinds of intelligence. Gardner has attempted to handle this problem by specifying as carefully as possible a set of criteria that must be met for something to be called an intelligence.

Sternberg's (1985) triarchic framework identifies three subtheories of intelligence. A componential subtheory of intelligence is concerned with the information-processing components that underlie intelligent performance. An experiential subtheory of intelligence is concerned with the roles of novelty and automatization in intelligent performance. A contextual subtheory of intelligence is concerned with selecting, shaping, and adapting to environments. It would appear that each of the subtheories could be applied to both academic and practical tasks.

There is very little empirical work that explicitly addresses the issue of relations between academic and practical intelligence. Given the recency of much of the practical intelligence work, this state of affairs is not surprising. Until more work is done, most researchers accept the consensus view that academic and practical intelligence represent partially overlapping rather than mutually exclusive aspects of intelligence (Ford, 1986).

The Case of Induction

Reasoning problems are commonly divided into those that require deduction and those that require induction (Sternberg, 1982). Deduction involves applying a principle to examples (deducing conclusions). The most commonly studied deduction problem is the syllogism. An example of a linear is, "Jim is taller than Mary; Mary is taller than Fred; who is tallest?"

Induction involves inducing a principle from examples. Induction tasks that have been examined include (a) the analogy, "doctor is to patient as lawyer is to ______"; (b) the series completion, "2, 6, 10, ______"; and (c) various classification and categorization tasks. An example of a classification task is the concept identification task in which a subject is presented with pairs of geometric objects such as large and small circles, squares, and triangles, and guesses which one represents the category. Feedback is provided, and over trials, subjects discover that the underlying concept is "small," "left," and so on.

It is difficult to consider what life would be like if we were not able to abstract rules and concepts effortlessly. The importance of induction is reflected by the attention it has been given by a number of theorists. Thurstone (1938) identified inductive reasoning as one of seven primary mental abilities, and induction is central to Cattell's concept of fluid abilities. Induction items comprise over half of the items on many IQ tests, and induction tasks have been a favorite of information-processing theorists (Pelligrino & Glaser, 1980; Sternberg, 1977a,b). Although induction appears central to our daily intellectual lives, the content of virtually all inductive reasoning tasks that have been studied to date has been academic and symbolic (e.g., red square).

Ethel Parker and I recently developed an induction task with content drawn from the social domain (Parker & Wagner, 1988). In contrast to the sparse and academic content of many induction tasks, the content of our task consisted of fictionalized episodes in a child's life. There were 10 problems. Each problem consisted of a set of seven episodes or trials, in which the main character of the story was faced with a choice. Unbeknownst to our subjects, there was a defining behavioral characteristic associated with each problem that governed the choice the main character would make. The defining behavioral characteristics we used included extrovert, quitter, sensitive, optimistic, introvert, irresponsible/forgetful, adventurous, aggressive, ambitious, and self-centered. Each trial consisted of a brief description of the main character in a situation, concluding with a choice the main character had to make. Consider the following example from "aggressive" Jane's story:

> The next day, Jane went to the store to spend some of her allowance. There were a lot of neat things and Jane had a hard time deciding what to buy. She was trying to decide between a Rambo-type squirt gun and a set of paints. Which one do you think she chose, the squirt gun or the paints?

A comprehensive check was made to insure that the subjects understood and remembered the information from the trial, and then the subjects were asked to predict what Jane would do and why, and to indicate what they would do and why. We asked subjects to indicate what they would do and why to test the hypothesis that children use self-knowledge to figure out the behavior of others. After the first trial, each subsequent trial began with a statement that indicated the choice Jane had made in the previous situation. This feedback was thus available for subjects to use to improve their predictive performance.

The subjects were fourth-grade students, seventh-grade students, and college undergraduates. We originally attempted to include a group of first grade students, but they were unable to perform at above chance level on the problems. In addition to the social induction task, we gave our subjects IQ tests to be able to determine the extent to which academic intelligence accounted for performance on the social induction task.

To conserve space, discussion will be limited to the prediction score results. There were 10 problems, each with 7 trials. A point was given for each correct prediction, yielding a total possible score of 70, with chance prediction being represented by a score of 35. The fourth-grade scores ranged from 33 to 57, with a mean of 42.8 and a standard deviation of 6.8. The seventh-grade scores ranged from 37 to 64, with a mean of 51.3 and a standard deviation of 8.5. The undergraduate scores ranged from 54 to 66, with a mean of 60.9 and a standard deviation of 2.9. The performance of the groups is plotted by trial position in Figure 11.1. A group-by-trial repeated measures analysis of variance with polynomial contrasts (linear and quadratic) on the repeated measures factor of trial yielded a main effect of groups, $F(2, 85) = 63.3, p < .001$, and both linear, $F(1, 85) = 359.6, p < .001$, and quadratic, $F(1, 85) = 73.5, p < .001$ effects of trial. There also were group-by-linear trial, $F(2, 85) = 28.6, p < .001$, and group-by-quadratic trial, $F(2, 85) = 28.6, p < .001$, and group-by-quadratic trial, $F(2, 85) = 13.7, p < .001$ interactions. Consider each effect in turn, referring to Figure 11.1.

The effect of group indicates that performance improves across developmental level. Follow-up analysis indicated that all three means were reliably different from each other. The linear effect of trial position indicated that performance improved across trials, and the quadratic effect of trial indicates that the improvement in performance was greater for earlier than later trials. The interactions of group with both the linear and quadratic effects of trial indicate that the improvement across trials, and especially early trials, is greater for older as opposed to younger subjects. These results are very orderly, with sug-

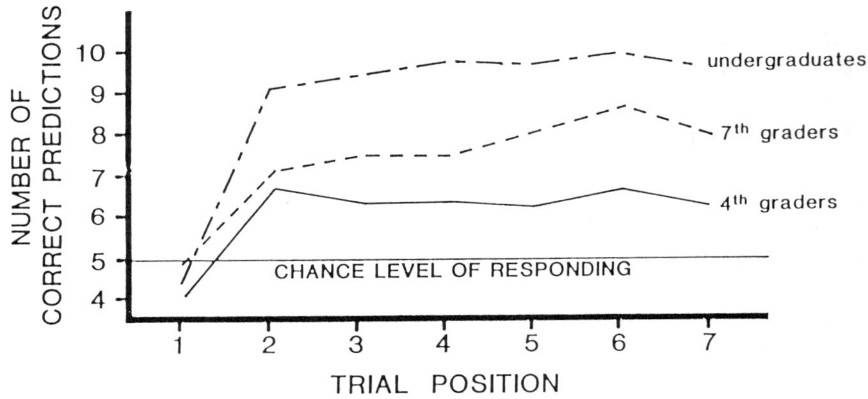

Figure 11.1. Number of correct predictions (out of 10 possible) across problems by trial position for the three groups.

gests that the task is measuring something other than junk. The reliabilities were satisfactory for the 4th and 7th graders, but perhaps due to ceiling effects, a bit low for undergraduates, with coefficient alphas of .68, .78, and .42, respectively.

The question of interest is what are the correlations between IQ and performance on the social induction task for the three groups. None of the correlations were reliably different from 0, with values of $-.02$ ($p > .05$) for fourth graders, .16 ($p > .05$) for seventh graders, and $-.13$ ($p > .05$) for undergraduates. Although the undergraduate results might be challenged by those who believe that IQ predicts performance in everyday situations on the grounds that undergraduates represent a restricted range of IQ and task performance was not very reliable, the fourth and seventh grade data are difficult to explain away. They suggest that induction in a social domain is little if at all related to academic intelligence.

Summary

In this section we have seen how researchers are moving beyond academic-type tasks in their analyses of intelligent performance, and thus are attempting to address one of the two limitations identified for the psychometric, Piagetian, and information-processing conceptions of intelligence. The concept of practical intelligence has figured prominently in this work. Several working definitions of practical intelligence were described, and proposed relations between academic and practical intelligence were examined. Results from a study of induction in a social context suggested that previous results derived from studies of IQ-type induction problems may not apply to induction in the everyday world. In the final section of the chapter, I will make some observations about where intelligence research appears to be headed in the near future.

THE FUTURE: LEARNING AGAIN?

As we have just seen, intelligence researchers have begun to address one of two limitations of prior work, that of the narrow range of academic-type tasks that have been studied. Addressing the second limitation, that research on intelligence has not dealt adequately with learning and education, is a likely agenda for intelligence research in the near future.

From the appearance of well-articulated theories of learning in the 1940s, a close relationship between intelligence and learning has been assumed (Estes, 1982). Today, there is little doubt that learning cannot proceed independently of intelligence (Estes, 1982), and conversely, that significant questions for the field of intelligence necessarily involve questions about the nature of learning (Glaser, 1986). There is a growing acceptance of the view that "learning and intelligence must be studied as continuously interacting components of the individual's whole cognitive system" (Estes, 1982, p. 191).

The assumption of a close relationship between intelligence and learning is reflected in the fact an individual's ability to learn is often viewed as central to intelligence. In a 1921 symposium on "Intelligence and Its Measurement," published by the *Journal of Educational Psychology*, a number of leading researchers of the day conceptualized intelligence in terms of learning ability. Colvin viewed intelligence as "the capacity of learn," and similarly, Dearborn viewed intelligence as "the capacity to learn or to profit from experience." Learning ability also figured prominently in the views of Henmon and Woodrow. More recently, Gagne (1968) has identified intelligence with the cumulative products of learning, and learning appears as a prominent feature in a number of recent multifaceted theories of intelligence—either in terms of the processes by which knowledge is acquired (Sternberg, 1985), or in terms of the knowledge itself (Glaser, 1986; see also Brown & Campione, 1986; Butterfield, 1986; Snow, 1986).

Historically, conceptualizing intelligence in terms of learning ability has generated considerable empirical activity, the results of which provide surprisingly little support for a close relationship between intelligence and learning. Beginning with Spearman (1904), a long series of studies ensued that examined correlations between performance on laboratory learning tasks and psychometric intelligence tests. The upshot of these studies is that there is a reliable but modest (.2-.4) relation between performance on laboratory learning tasks and intelligence, falling far short of what would be expected if there indeed was a close relationship between intelligence and learning (Cronbach & Snow, 1977; Estes, 1982; Snow & Yalow, 1982; Woodrow, 1940; Zeaman & House, 1967).

On both theoretical and logical grounds, it would appear that there must be important relations between intelligence and learning, but, after decades of study of both intelligence and learning, the nature of these relations remains a perplexing mystery (Butterfield, 1986; Estes, 1982). Why, after so much effort, is there so little understanding of the relationship between intelligence and learning?

Part of the problem may be that learning and intelligence began and evolved as separate disciplines, a separation perpetuated by the fact that most new research grows directly from old research (Estes, 1982). Another part of the problem may be that, whereas conceptions and measurement operations of both learning and intelligence have been extraordinarily useful for some purposes, the same conceptions and measurement operations have been less useful for other purposes, including understanding how intelligence and learning are related. Learning research has too often been based on the unfounded assumption that the principles of learning can be studied in their entirety using highly simplified learning tasks (Estes, 1982); intelligence research has too often been based on the unfounded assumption that intelligence, as measured by traditional psychometric tests, is equivalent to intelligence as it operates in the everyday world (Neisser, 1976; Glaser, 1986; Wagner & Sternberg, 1985).

Two relatively recent developments, which occurred simultaneously in the fields of intelligence and learning, have implications for understanding relationships between intelligence and learning: The first development is increased attention to the study of performance in more realistic situations; the second, concomitant, development is increased attention to the role of prior knowledge in performance.

Turning first to the field of learning, there is a growing body of learning research that is based on more realistic tasks, including those we might characterize as school learning (Brown, 1982; Greeno; 1980; Siegler, 1983; Siegler & Klahr, 1982; Snow & Yalow, 1982, Voss, 1978). It is likely that performance on such tasks is more closely related to intelligence than is performance on simplified laboratory tasks (Estes, 1982). For example, typical correlations between IQ scores and various indices of school learning (.4 to .7), are roughly double the correlations between IQ and performance on laboratory learning tasks.

Increased attention to the role of prior knowledge in learning is evident in the view that more of what remains to be learned about learning—including further understanding of basic principles—will come from study of interactions of new information with existing knowledge structures, rather than from further study of tasks that minimize the role of prior knowledge in performance (Greeno, 1980; Siegler, 1983).

Turning to the field of intelligence, the growing body of research on intelligence as it operates in the everyday world was mentioned in the previous section of this chapter (see Sternberg & Wagner, 1986, for a collection of recent approaches).

Increased attention to the role of prior knowledge in intelligent performance is evident by the interest in the study of expert–novice differences in a variety of domains (see Chi, Glaser, & Rees, 1982, for a review of this literature). An important conclusion from the study of expert–novice differences is that experts differ from novices primarily in their knowledge than can be brought to bear on a task, and in how this knowledge is structured, rather than in their underlying cognitive abilities. In addition, the acquisition and use of practical knowledge has figured prominently in accounts of intelligence as it operates in the everyday world (Berry & Irvine, 1986; Wagner, 1987; Wagner & Sternberg, 1985).

Summary

On both theoretical and logical grounds, learning and intelligence are closely intertwined. That so little progress has been made in specifying the role of learning in intelligence and that of intelligence in learning is understandable, given that the learning tasks chosen for study in the laboratory were selected for their impoverished content and that the field of intelligence has traditionally been too enamored with measurement to the neglect of a theoretical foundation that reflects what is known about cognition. The increased interest in more realistic learning tasks, and in intelligence as it operates in the everyday world, has brought the issue of the acquisition and use of knowledge to the forefront. This topic is likely to dominate new work in intelligence in the coming years, the goal of which will be a theory of intelligence with real implications for learning.

CONCLUSIONS

This chapter examined past and present conceptions of intelligence, and suggested likely directions for future conceptions. The three conceptions of intelligence have dominated the field through the mid-1970s—the psychometric, the Piagetian, and the information-processing—are complementary in that they address three different aspects of performance. The psychometric conception of intelligence has sought to account for differences in performance across individuals (subject variance). The information-processing conception of intelligence has sought to account for differences in performance across items (stimulus variance). The Piagetian conception of intelligence has sought to account for developmental differences in performance. The two basic limitations of these conceptions are that their support derives from an empirical base that is limited to a very narrow range of academic-type tasks, and their relations with learning are poorly specified.

Recent work has addressed the limitation of an empirical base that is limited to a narrow range of academic-type tasks, in that researchers have moved beyond IQ and academic-type tasks in their analyses of intelligent performance. The concept of practical intelligence has figured prominently in this work. Results from a study of induction in a social context suggests that previous results derived from studies of IQ-type induction problems may not apply to induction in the everyday world.

Addressing the role of poorly specified relations between intelligence and learning is likely to figure prominently on the agenda for research on intelligence in the near future. That so little progress has been made in specifying the role of learning in intelligence, and that of intelligence in learning, is a consequence of choosing learning tasks for study in the laboratory that had impoverished content, and of the field of intelligence having been too enamored with measurement issues to the neglect of a modern theoretical foundation. As learning researchers and intelligence researchers have become interested in more realistic tasks, the issue of the acquisition and use of knowledge has emerged as a critical one.

There are a number of implications of this examination of conceptions of intelligence for researchers interested in the intergenerational transfer of cognitive skills, including the cognitive skills that represent literacy.

First, real competency involves being able to use one's cognitive skills with little outside prompting and in diverse contexts. Thus, when deciding on the nature of cognitive skills that one should study in designing studies of intergenerational transfer of cognitive skills, it is important that cognitive skills be broadly rather than narrowly defined and that practical as well as academic aspects are addressed.

Second, prior to attempts to intervene in the intergenerational transfer of cognitive skills, it is important to understand the nature of the informal learning that takes place between parent and child. One potentially valuable source of ideas for intervention is studies of individual differences, within subcultural settings and perhaps between as well, in the extent to which children learn from their parents. Such studies should distinguish between individual differences in the cognitive skills that parents have to transfer to children, and individual differences in the parents' ability and willingness to do so.

Third, the potentially instrumental role of the child in intergenerational transfer should not be ignored; it is amazing how ingenious some children can be when attempting to get what they want from their parents. It also is the case that children who ask parents to teach them to fish or cook or play ball "like mommy or daddy does" are motivated to learn in a way that is not often seen in school. The exciting possibility is that this heightened motivation for learning might be used to transfer cognitive skills such as literacy as well.

REFERENCES

Atkinson, R. L., Atkinson, R. C., Smith, E. E., & Hilgard, E. R. (1987). *Introduction to psychology*. New York: Harcourt Brace Jovanovich.

Berry, J. W., & Irvine, S. H. (1986). Bricolage: Savages do it daily. In R. Sternberg & R. Wagner (Eds.), *Practical intelligence* (pp. 271-306). New York: Cambridge University Press.

Binet, A., & Simon, T. (1905). Methodes nouvelles pour le diagnostic du niveau inellectuel des anormaux. *L'Annee Psychologique, 11*, 191-244.

Brainerd, C. J. (1978). The stage question in cognitive-developmental theory. *Behavioral and Brain Sciences, 1*, 173-182.

Brown, A. L. (1982). Learning and development: The problems of compatibility, access, and induction. *Human Development, 25*, 89-115.

Brown, A. L., & Campione, J. C. (1986). Academic intelligence and learning potential. In R. Sternberg & D. Detterman (Eds.), *What is intelligence* (pp. 39-43). Norwood, NJ: Ablex.

Brown, G., & Desforges, C. (1979). *Piaget's theory: A psychological critique*. Boston: Routledge & Kegan Paul.

Butterfield, E. C. (1986). Intelligent action, learning and cognitive development might all be explained with the same theory. In R. Sternberg & D. Detterman (Eds.), *What is intelligence?* (pp. 45-49). Norwood, NJ: Ablex.

Chi, M. T. H., Glaser, R., & Rees, E. (1982). Expertise in problem solving. In R. Sternberg (Ed.), *Advances in the psychology of human intelligence* (Vol. 1, pp. 7-75). Hillsdale, NJ: Erlbaum.

Cronbach, L. J., & Snow, R. E. (1977). *Aptitudes and instructional methods.* New York: Irvington.

Eisenberg, L., & Conners, C. K. (1966). *The effect of Head Start on developmental process.* Paper presented at the 1966 Joseph P. Kennedy, Jr. Foundation Scientific Symposium on Mental Retardation, Boston, MA.

Estes, W. K. (1982). Learning, memory, and intelligence. In R. Sternberg (Ed.), *Handbook of human intelligence* (pp. 170-221). New York: Cambridge University Press.

Ford, M. E. (1986). For all practical purposes: Criteria for defining and evaluating practical intelligence. In R. Sternberg & R. Wagner (Eds.), *Practical intelligence* (pp. 183-200). New York: Cambridge University Press.

Ford, M. E., & Miura, I. (1983). *Children's and adults' conceptions of social competence.* Paper presented at the meeting of the American Psychological Association, Anaheim, CA.

Frederiksen, N. (1986). Toward a broader conception of human intelligence. In R. Sternberg & R. Wagner (Eds.), *Practical intelligence* (pp. 84-116). New York: Cambridge University Press.

Furth, H. G. (1970). *Piaget for teachers.* Englewood Cliffs, NJ: Prentice-Hall.

Gagne, R. M. (1968). Contributions of learning to human development. *Psychological Review, 75,* 177-191.

Galton, F. (1888). Co-relations and their measurements, chiefly from anthropometric data. *Proceedings of the Royal Society, 45,* 135-140.

Gardner, H. (1983). *Frames of mind.* New York: Wiley.

Glaser, R. (1986). Intelligence as acquired proficiency. In R. Sternberg & D. Detterman (Eds.), *What is intelligence?* (pp. 77-83). Norwood, NJ: Ablex.

Greeno, J. G. (1980). Some examples of cognitive task analysis with instructional implications. In R. Snow, P. Federico, & W. Montague (Eds.), *Aptitude, learning, and instruction: Vol. 2.* Hillsdale, NJ: Erlbaum.

Mercer, J. R., Gomez-Palacio, M., & Padilla, E. (1986). In R. Sternberg & R. Wagner (Ed.), *Practical Intelligence* (pp. 307-337). New York: Cambridge University Press.

Neisser, U. (1976). General, academic, and artificial intelligence. In L. Resnick (Ed.), *The nature of intelligence* (pp. 135-144). Hillsdale, NJ: Erlbaum.

Newell, A., & Simon, H. A. (1972). *Human problem solving.* Englewood Cliffs, NJ: Prentice-Hall.

Parker, E., & Wagner, R. K. (1988). *Induction in a social domain.* Manuscript submitted for publication.

Pearson, K. (1901). On lines and planes of closest fit to systems of points in space. *Philosophical Magazine, 6,* 559-572.

Pelligrino, J. W., & Glaser, R. (1980). Components of inductive reasoning. In R. E. Snow, P. A. Federico, & W. E. Montague (Eds.), *Aptitude, learning, and instruction: Vol. 1. Cognitive process analysis of aptitude* (pp. 177-217). Hillsdale, NJ: Erlbaum.

Penrose, L. S. (1963). *The biology of mental defect.* London: Sidgwick & Jackson.

Sarason, S. B. (1953). *Psychological problems in mental deficiency.* New York: Harper.

Siegel, L. S., & Brainerd, C. J. (Eds.). (1978). *Alternatives to Piaget: Critical essays on the theory.* New York: Academic Press.

Siegler, R. S. (1983). Five generalizations about cognitive development. *American Psychologist, 38*, 263-277.

Siegler, R. S., & Klahr, D. (1982). When do children learn? The relationship between existing knowledge and the ability to acquire new knowledge. In R. Glaser (Ed.), *Advances in instructional psychology.* Hillsdale, NJ: Erlbaum.

Snow, R. E. (1978). Theory and method for research on aptitude processes. *Intelligence, 2*, 225-278.

Snow, R. E. (1986). On intelligence. In R. Sternberg & D. Detterman (Eds.). *What is intelligence?* (pp. 133-139). Norwood, NJ: Ablex.

Snow, R. E., & Yalow, (1982). Education and intelligence. In R. Sternberg (Ed.), *Handbook of Human Intelligence* (pp. 493-585). New York: Cambridge University Press.

Spearman, C. E. (1904). "General intelligence," objectively determined and measured. *American Journal of Psychology, 15*, 201-292.

Sternberg, R. J. (1977a). *Intelligence, information processing, and analogical reasoning: The componential analysis of human abilities.* Hillsdale, NJ: Erlbaum.

Sternberg, R. J. (1977b). Component processes in analogical reasoning. *Psychological Review, 84*, 353-378.

Sternberg, R. J. (1980). Sketch of a componential theory of human intelligence. *Behavioral and Brain Sciences, 3*, 573-584.

Sternberg, R. J. (1982). A componential approach to intellectual development. In R. Sternberg (Ed.), *Advances in the psychology of human intelligence* (Vol. 1). Hillsdale, NJ: Erlbaum.

Sternberg, R. J. (1985). *Beyond IQ: A triarchic theory of human intelligence.* New York: Cambridge University Press.

Sternberg, R. J., & Berg, C. A. (1986). Definitions of intelligence: A comparison of the 1921 and 1986 symposia. In R. Sternberg & D. Detterman (Eds.), *What is intelligence?* (pp. 155-162). Norwood, NJ: Ablex.

Sternberg, R. J., & Wagner, R. K. (Eds.) (1986). *Practical intelligence: Nature and origins of competence in the everyday world.* New York: Cambridge University Press.

Terman, L. M., & Merrill, M. A. (1960). *Stanford-Binet intelligence scale.* Boston: Houghton-Mifflin.

Thurstone, L. L. (1938). *Primary mental abilities.* Chicago: University of Chicago Press.

Tizard, J. (1953). The prevalence of mental subnormality. *Bulletin of the World Health Organization, 9*, 423-440.

Vernon, P. E. (1971). *The structure of human abilities.* London: Methuen.

Voss, J. F. (1978). Cognition and instruction: Toward a cognitive theory of learning. In A. Lesgold, J. Pellegrino, S. Fokkema, & R. Glaser (Eds.), *Cognitive psychology and instruction*. New York: Plenum Press.

Wagner, R. K. (1987). Tacit knowledge in everyday intelligent behavior. *Journal of Personality and Social Psychology, 52*, 1236-1247.

Wagner, R. K., & Kistner, J. A. (in press). Implications of the distinction between academic and practical intelligence for learning-disabled children. In H. L. Swanson & B. Keogh (Eds.), *Learning disbilities: Theoretical and research issues*. Hillsdale, NJ: Erlbaum.

Wagner, R. K. & Sternberg, R. J. (1984). Alternative conceptions of intelligence and their implications for education. *Review of Educational Research, 54*, 179-223.

Wagner, R. K., & Sternberg, R. J. (1985). Practical intelligence in real-world pursuits: The role of tacit knowledge. *Journal of Personality and Social Psychology, 48*, 436-458.

Windle, C. (1966). Prognosis of mental subnormals. *American Journal of Mental Deficiency, 66* (Monograph Supplement to No. 5).

Woodrow, H. (1940). Interrelations of measures of learning. *Journal of Psychology, 10*, 49-73.

Zeaman, D., & House, B. J. (1967). The relation of IQ and learning. In R. Gagne (Ed.), *Learning and individual differences*. Columbus, OH: Merrill.

Zigler, E., & Seitz, V. (1982). Social policy and intelligence, In R. Sternberg (Ed.), *Handbook of human intelligence* (pp. 586-641). New York: Cambridge University Press.

Chapter 12

Generalizability and Practical Intelligence*

Barbara Means
SRI International

The recent burgeoning of studies of "practical intelligence"—as exemplified by work like Parker and Wagner's (this volume) investigation of social cognition, or Schliemann and Acioly's (1989) study of the use of mathematical procedures by Brazilian bookies—makes us realize just how narrow traditional concepts of intelligence had become. As Neisser (1976) has argued, academics created the concept of intelligence in their own image, and the qualities that lead to success in school are a far cry from those that produce adaptive behavior in environments that have little in common with schools. (Neisser uses a battlefield example). Studies such as Wagner and Sternberg's (1985) research on business managers and Ceci and Liker's (1986) investigation of racetrack handicappers not only clarify the nature of intelligent behavior in everyday settings, but also document the lack of a relationship between individual differences in intelligent functioning in these settings and standard IQ measures.

* This work was supported by Air Force Contract No. F33615-84-C-0058 to the Human Resources Research Organization. The methods and work described here reflect the contributions of the many organizations and individuals involved in that project, including Chris Roth (HumRRO), Sherrie Gott and Bob Pokorny (Air Force Human Resources Laboratory), Allan Collins (BBN Laboratories), and Alan Lesgold and Gary Eggan (Learning Research and Development Center).

Although most of us have the sense that we "know it when we see it," practical intelligence remains a concept in search of a definition. In some treatments (e.g., Mercer, 1978), practical intelligence is characterized as adaptation to a social or cultural environment. This work reminds us that success in real-world settings draws heavily upon the individual's ability to adapt to a set of *social* roles and expectations. Although Wagner and Sternberg come from a tradition different from that of most of the social adaptation researchers, their measures of the "tacit knowledge" required for success in a domain are, to a large extent, standardized assessments of an individual's understanding of the real goals and avenues toward success within a particular occupational culture.

Other work on practical intelligence, such as Scribner's (1984) study of dairy workers and Ceci and Liker's (1986) study of racetrack handicappers, focuses more on intelligent grappling with task requirements within a particular work setting. This work bears a similarity to the growing literature on expertise in a variety of domains (see Chi, Glaser, & Rees, 1982) and stresses the application of specialized knowledge.

Another approach discussed by Wagner defines the boundaries of practical intelligence as competence in dealing with any ill-defined, open-ended problem with multiple solutions—provided that the problem is not academic in nature. It is interesting to note the resemblance to some definitions of higher-order thinking skills (except for the stipulation that the problem cannot be academic). Exemplified by Frederiksen's (1962) in-basket test for government administrators, which identified factors such as Preparing for Action by Becoming Informed and Directing Subordinates, this work puts less emphasis on domain-specific knowledge than does the expert–novice research.

Although disparate in foci, each of these definitions taps important aspects of what we mean when we characterize someone as practically intelligent, and each is associated with a valuable line of research. I would like to see all of these research efforts continue, and, therefore, see little value in trying to arrive at a single definition of practical intelligence. In fact, I would like to propose yet another definition in the hopes of similarly stimulating further work.

My motivation for proposing yet another approach toward conceptualizing practical intelligence stems from work performed for the Air Force while I was at the Human Resources Research Organization. After studying electronics technicians in three different Air Force jobs, I became convinced that a critical marker of job expertise—or practical intelligence—is representing job problems in a way that fosters generalization across problems and across settings. It is now well documented that expert performance in a particular domain is characterized by a more extensive, better organized knowledge base (Chase & Simon, 1973; Feltovich, 1981; Jeffries, Turner, Polson, & Atwood, 1981). But I would argue that intelligence in the workplace encompasses not just extensive knowledge relevant to the job as customarily practiced, but also knowledge that enables the individual to cope with unexpected problems and with significant changes in the job (e.g., the introduction of automation or a new system). In the next section of this chapter I will illustrate what I mean by "problem representations that foster generalization" and then draw some inferences about the utility of this perspective for studying individual differences in practical intelligence.

EXPERTISE AMONG ELECTRONICS TECHNICIANS

The purpose of our Air Force research was to identify the skills, knowledge, and strategies used by proficient electronics technicians as a foundation for developing assessment instruments and training systems. Our strategy for identifying job skills required going to Air Force bases to work intensively with the best maintenance personnel for a given electronics system at that base. We asked experts, first, to describe the system they maintained and the hard aspects of their job. These turned out to involve particular types of troubleshooting, specifically cases in which built-in tests and standard procedures are unsuccessful. After providing exhaustive lists of types of faults that are difficult to troubleshoot, experts and project researchers worked together to categorize the faults into sets of related problem types for further study. Experts were then asked to develop specific troubleshooting problems (hypothetical faults) representing each category.

We were able to observe troubleshooting reasoning by having the expert who invented a problem and the researcher who documented it team up to give the problem to another expert, who had not been told where the fault was or what kind of problem was involved. Given only the initial problem symptoms, just as he would encounter them in the shop, the expert problem solver was asked to describe how he would troubleshoot, thinking aloud as he described each step. The researcher recorded each action and also the goal or hypothesis motivating it, the result of the action (e.g., a voltage reading), and the problem solver's interpretation of the result. (See Means & Gott, 1988, for a fuller treatment of this method.)

One practical problem we encountered in performing this research was identifying *experts*. At each base, we asked the supervisor of the maintenance shop to identify the best available technicians—whether military or civilian—for the system under study. Initially we asked for six *experts*, but soon learned that, given the relatively small size of many maintenance shops (e.g., fewer than 50 people), this request usually netted two or three truly expert technicians, several good ones, and one or two who would have to be regarded as less than expert. Moreover, we found no ready method for separating the truly expert from the less than expert in advance of interviewing them in some depth. Years on the job appeared to be a necessary factor for expertise but not a sufficient one. Some technicians with 5 to 10 years of experience were quite good, but others had many weaknesses. (The very best experts tended to be civilians who were former enlisted personnel and had more than 10 years' experience.) Test scores were not available for many candidate experts and held little promise anyway since, in a related, earlier project, Glaser et al. (1985) found no relationship between troubleshooting skill and either a general measure of aptitude (the Armed Forces Qualification Test score) or paper-and-pencil tests of electronics knowledge.

Although difficult to index in advance, the individual differences we observed in our work with a range of electronics maintenance personnel were striking, not only in terms of contrasts in ability to solve problems efficiently, but also in the way individuals structured their solutions and in the way they talked about the system they maintained. After working with a number of different technicians in three different jobs, I formed some general ideas concerning the source of these differences: Expert technicians appear to use an abstract functional model of the system on which they work as an aid in categorizing problem symptoms and selecting hypotheses. When troubleshooting complex systems such as those we studied, this ability to categorize problems in order to narrow the search space is absolutely critical. The importance of having and using such abstract mental models can be illustrated with a contrasting pair of problem solutions.

AN ILLUSTRATION OF INDIVIDUAL DIFFERENCES

Both sample solutions involve the same troubleshooting problem, developed for a job in an intermediate maintenance shop for the F-15 aircraft. To understand the examples, a general overview of the technician's job is useful. When portions of the aircraft require routine maintenance or are suspected of malfunction, they are pulled from the aircraft on the flightline and sent to the intermediate maintenance shop for testing and repair. In the shop, the black boxes are hooked up to an automatic test station, that simulates the aircraft, sending signals into the black box and then receiving the return signal and comparing it to acceptable values for that test. Interestingly, the hardest part of this job is troubleshooting faults in the test station, which consists of a central computer and some 100 cubic feet of electronic circuitry for generating, conditioning, and measuring signals. (The automated diagnostic software is pretty good at identifying the cause for various faults in the black boxes but much less reliable for the test station itself.) In this problem, the hypothetical technician has come into the shop and tries to turn on the test station to begin work, but finds that, when he turns the system on, the front panel lights on the main computer start flashing on and off at random, and that he is unable to load any programs.

Given these symptoms, one technician (with over 5 years' experience) began by turning to the system technical documentation, looking for something connected to the front panel lights on the computer section of the test station. The problem with this approach is that the technical documentation for maintaining this system comprises more than 60 volumes. The technician went leafing through page after page, looking for "where it's getting its power supply, or its ground, or its control from." After 45 minutes he avowed that he had no idea what the problem was.

This fruitless, undirected search can be contrasted to the approach of a really expert technician. Given the same symptoms, his first reaction was not to dive into detailed technical documentation, but rather to reason about the type of problem indicated by the given symptom. All such systems involve power, signal, data, routing, and measurement. Such massive widespread symptoms suggest a problem with either power or data (computer control). Since the former type of problem is easier to isolate, that hypothesis should be pursued first. This tentative categorization of problem symptoms in terms of relevant system functions turns out to have a profound effect upon problem solving. Instead of looking for *anything* connected to the front panel lights, the expert limits his search to the three sources of power to the main computer. Thus, the only documentation he needs to look at to refine his hypothesis is a single-page power distribution diagram which shows the input pins and voltages for power from the three sources. Testing these inputs, he quickly isolates the problem to one of the three power sources (and proceeds to make tests to identify the bad part within the major component).

Although there were many differences between the two problem solvers, I would argue that the abstract system representation and use of that representation to classify symptoms is a crucial part of the expert's ability to organize his search through the space of possible fault locations. This reference to a small set of system functions during hypothesis formation was found repeatedly among the best problem solvers in the three jobs we studied.

In addition to using such abstract functional models to organize their problem solving, experts also used this kind of organization when asked to describe the system to us. They would talk about computer control within the system, discussing the roles of relevant components and then moving to the description of the parts of the system involved in signal generation, components involved in routing signals, and so on. In contrast, less-expert technicians were likely to organize their descriptions by the physical location of the components, describing the contents and functions of the top lefthand major component, then the component just below it, and so on. The basic argument is that, in both cases, technicians are acquiring job-specific knowledge, but that the technician who relates that knowledge to a more generalizable, abstract model will have an advantage both in working with unusual problems arising in the current job and in moving to a new or modified job.

IMPLICATIONS OF THESE DIFFERENCES

There are several encouraging aspects about the view that practical expertise takes this form:

1. Such abstract, functional system representations should be teachable. Although the kinds of simplifying, functional models experts use are not routinely taught at present, there is nothing about the models to suggest that they are too abstract or too complex to be learned by most technicians. This approach will be tried out in two training systems now under development at BBN Laboratories (Kurland & Tenney, 1988; White & Frederiksen, 1988).
2. The functional system representations appear general enough to apply across a range of jobs. We found that the same system representation applied to two different jobs involving automated test stations, and that a very similar model was used in the telecommunications system maintenance job that we studied.

The latter fact is at the crux of my argument for studying problem representation and generalization as aspects of practical intelligence. Figure 12.1 shows the mapping between a single abstract functional

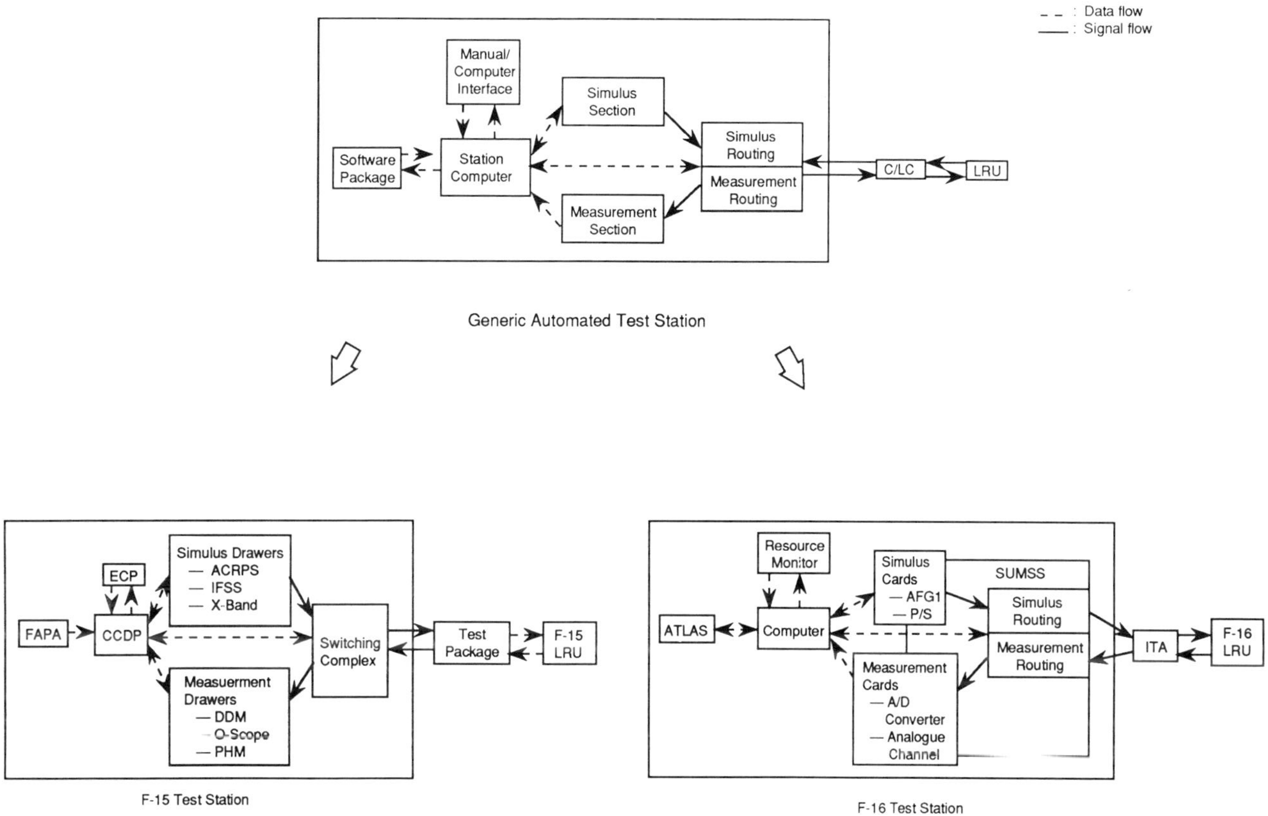

Figure 12.1. Relationship between an abstract, functional model and two specific automatic test stations.

model and the physical models for two different systems (automatic test stations for the F-15 and F-16 aircraft, respectively). This mapping demonstrates that generalization of a functional model across systems is possible in principle. Further, the protocols provided by technicians who had maintained a number of different systems suggest that this abstract functional system model is something that good technicians bring from one job to another. Although each system is physically different and has its own peculiarities in terms of the particular components that tend to go bad, the more abstract model, and associated categories of problems, still apply. Our own introspective experience while conducting this research was that we used the abstract model we learned, in working with technicians from the first job we studied, to understand and organize what we learned about the second and third systems.

Good performance in a job draws upon both job-specific knowledge (e.g., the CCDP in the F-15 automatic test station is very reliable), and procedures (steps for calibrating a specific component) and upon the more generalizable knowledge characterized above as an abstract, functional system model. Although such models are not as widely applicable as "weak methods" or "common sense," my intuition is that such knowledge would have an advantage over more general knowledge structures in terms of their power to enhance job performance and in terms of trainability. Although not applicable to all jobs or all parts of life, such models are general enough that those who possess them should be better able to cope with changes in the system they maintain or with a new system. In short, they should exhibit practical intelligence in the workplace. Similar arguments could be constructed about the importance of generalizable models for functioning in other real-world settings.

In a sense, then, I am returning to the familiar notion that transfer is a critical part of intelligence, but here I am calling for psychologists to study transfer in real-life situations rather than in the experimental laboratory. I am suggesting that a critical aspect of practical intelligence is representing the problem in a way that promotes generalization. Some people seem to be more likely to do this than others, but it should be remembered that our present education and training systems do little to facilitate such transfer. With a careful analysis of the mental models used by experts in a set of related jobs, we should be able to produce really significant improvements in performance by training generalizable models that offer some real conceptual leverage when the individual is faced with variations in job demands. Thus, while heartily commending the work that has been done on practical intelligence thus far, I am calling for its expansion to include the study of real-world generalization.

REFERENCES

Ceci, S. J., & Liker, J. (1986). Academic and nonacademic intelligence: An experimental separation. In R. J. Sternberg & R. K. Wagner (Eds.), *Practical intelligence* (pp. 119-142). Cambridge, UK: Cambridge University Press.

Chase, W., & Simon, H. (1973). Perception in chess. *Cognitive Psychology, 1,* 55-81.

Chi, M. T. H., Glaser, R., & Rees, E. (1982). Expertise in problem solving. In R. J. Sternberg (Ed.), *Advances in the psychology of human intelligence* (Vol. 1, pp. 7-75). Hillsdale, NJ: Erlbaum.

Feltovich, P. J. (1981). *Knowledge-based components of expertise in medical diagnosis.* Unpublished doctoral dissertation, University of Minnesota.

Frederiksen, N. (1962). Factors in in-basket performance. *Psychological Monographs, 76* (22, No. 541).

Glaser, R., Lesgold, A., Lajoie, S., Eastman, R., Greenberg, L., Logan, D., Magone, M., Weiner, A., Wolf, R., & Yengo, L. (1985). *Cognitive task analysis to enhance technical skills training and assessment.* (Final Report to the Air Force Human Resources Laboratory on Contract No. F41689-83-C-0029). Pittsburgh: Learning Research and Development Center, University of Pittsburgh.

Jeffries, R., Turner, A. T., Polson, P. G., & Atwood, M. E. (1981). Processes involved in designing software. In J. Anderson (Ed.), *Cognitive skills and their acquisition* (pp. 255-283). Hillsdale, NJ: Erlbaum.

Kurland, L. C., & Tenney, Y. J. (1988). Issues in developing an intelligent tutor for a real-world domain. In J. Psotka, L. D. Massey, & S. A. Mutter (Eds.), *Intelligent tutoring systems: Lessons learned* (pp. 119-180). Hillsdale, NJ: Erlbaum.

Means, B., & Gott, S. P. (1988). Cognitive task analysis as a basis for tutor development. In J. Psotka, L. D. Massey, & S. A. Mutter (Eds.), *Intelligent tutoring systems: Lessons learned* (pp. 35-57). Hillsdale, NJ: Erlbaum.

Mercer, J. R. (1978). Theoretical constructs of adaptive behavior: Movement from a medical to a social-ecological perspective. In W. A. Coulter & H. W. Morrow (Eds.), *Adaptive behavior: Concepts and measurements.* New York: Grune & Stratton.

Neisser, U. (1976). General, academic, and artificial intelligence. In L. Resnick (Ed.), *The nature of intelligence* (pp. 135-144). Hillsdale, NJ: Erlbaum.

Schliemann, A. D., & Acioly, N. M. (1989). Mathematical knowledge developed at work: The contribution of practice versus the contribution of schooling. *Cognition and Instruction, 6,* 185-221.

Scribner, S. (Ed.). (1984). Cognitive studies of work [Special issue]. *Quarterly Newsletter of the Laboratory of Comparative Human Cognition, 6* (1&2).

Wagner, R. K., & Sternberg, R. J. (1985). Practical intelligence in real-world pursuits: The role of tacit knowledge. *Journal of Personality and Social Psychology, 48,* 436-458.

White, B. F., & Frederiksen, J. (1988, April). *Intelligent learning environments for science education: The need for multiple mental models.* Paper presented at the annual meeting of the American Educational Research Association, New Orleans.

Chapter 13

Getting the Message Across: Cultural Factors in the Intergenerational Transfer of Cognitive Skills

Stephen Reder
Northwest Regional Educational Laboratory

When first asked to discuss the cultural factors underlying the intergenerational transfer of cognitive skills, the scope of the assignment seemed too broad. After all, if we use one popular definition of *culture*—that which is learned and passed on between people and across generations—then cultural processes are a critical medium for the intergenerational transmission of cognitive skills. But the very breadth and complexity of these cultural processes requires the topic to be narrowed to be usefully treated in a single chapter.

There are, of course, well-established academic traditions for studying the cultural basis of intellectual development. We could narrow our focus by choosing one of these approaches. Some approaches interpret apparent differences in cognition between groups, cultures, or societies in terms of fundamental differences in underlying cognitive capabilities or skills (e.g., Goody & Watt, 1963; McLuhan, 1964; Havelock, 1963; Olson, 1977); other approaches focus on the attendant differences in the social and institutional organization of observable practices or behaviors (e.g., Scribner & Cole, 1981; Eisenstein, 1979; Leach, 1982; Goodman, 1977).

Without attempting to justify our choice by adding to an already heated debate over which is the more productive approach (Cole & Scribner, 1976; McDermott & Hood, 1982), an analytical framework will be used here which focuses on the *socialization* of specific types of *intellectual behavior*. Rather than looking at the cultural transmission of cognitive abilities, the focus thus shifts to the analysis of the settings, social interactions, and contexts in which the target behaviors take place and are socialized to others.

A socialization framework offers the possibility of integrating what often are separately studied environmental influences on cognitive development: school, home, and community. Social interactions are examined closely as contexts for the display and development of intellectual skills, regardless of whether the interactions are between teacher and student in the classroom, between parent and child in the home, between friends in recreational settings, between coworkers on the job, and so on. By attempting to identify and analyze the influence of multiple contexts and activity domains on the development of intellectual behaviors, our approach will be able to suggest a correspondingly richer set of policy and implimentation options. Reder and Green (1985), for example, applied such a framework to literacy development and urged educators, not only to consider strategies for changing the *individual* (e.g., by increasing someone's skill level), but to envision ways to change the individual's *environment* which might better facilitate the individual's literacy development.

Given the wide range of interactional contexts which potentially impact an individual's cognitive development, considerable care must be taken in identifying specific socialization processes with the "intergenerational transfer" of a particular cognitive skill or intellectual behavior. Some cognitive skills or behaviors may be socialized directly from parents to children in one group of families, for example, but subsequently socialized to other children through peer interactions in school or community settings. The role of formal education in the socialization of the intellect similarly may be expected to be complex and varied. Thus, issues and findings about the intergenerational transfer of cognitive skills must be placed within a broader socialization framework which incorporates multiple interactional contexts.

In this chapter, literacy development will be used as an example of the socialization of (the cognitive skills underlying) intellectual behaviors. Literacy is a useful choice because of the burgeoning multidisciplinary body of recent research about its development (e.g., Bloome, 1987; Olson, Torrance, & Hildyard, 1985; de Castell, Luke, & Egan, 1986; Wagner, 1987; Langer, 1987), particularly in relation to contextual and interactional influences (e.g., Scribner & Cole, 1981;

Heath, 1983; Reder & Green, 1983; Scollon & Scollon, 1981; Schiefflen & Gilmore, 1986; Cook-Gumperz & Gumperz, 1981; Reder, 1987). Literacy is also, of course, a realm of intellectual activity in which researchers, educators, and policy analysts perceive a problem whose solution requires an intergenerational approach (Hess & Holloway, 1979; Sticht, 1983). Progress made in understanding and facilitating the intergenerational transfer of literacy will contribute greatly to parallel efforts involving other cognitive skills.

The analysis of the socialization of literacy practices is organized into several sections. Part 2 develops an analytical framework for describing *what* it is that is socialized, that is, the nature and social organization of literacy practices. Part 3 considers the question of *how* such literacy practices are socialized. The final section considers the implications of these sections for educational policy, practice, and research.

THE CULTURAL ORGANIZATION OF LITERACY PRACTICES

Literacy and Social Interaction

Clifford Geertz, in his seminal book *Interpretation of Cultures* (1973), defines culture as the "webs" of meanings and interpretations that are created through social interactions. These meanings and interpretations are not to be understood as knowledge that is carried around in peoples' heads, but as an ongoing dynamic process which is created and recreated through the actions individuals take in interacting with each other, and the interpretations and meanings they create and share through those interactions.

Geertz's view of culture is a very useful departure point for analyzing the cultural transmission of literacy. If *literacy* is now taken to be the use of written language in cultural activities (whatever those activities may be), rather than a set of underlying cognitive capabilities, the analysis focuses on exploring reading and writing activities as cultural *events*. The critical issues for the cultural transmission of literacy concern how people, in the course of their interactions, create and recreate contexts that influence and are influenced by the use of written language, and how the text and actions involved in those interactions are interpreted and given meaning by the participants.

The application of this perspective to literacy research entails careful observation and analysis of naturally occurring literacy events, that is, interactions in which individuals utilize written materials. Such interactions include, but are not limited to, those events in which only one individual encounters written materials; many, if not most, naturally occurring literacy events involve social interactions between cooperating individuals as well as interactions between individuals and written materials (Heath, 1983; Fingeret, 1983; Taylor, 1983; Reder & Green, 1983, 1985).

Literacy Practices

Literacy events are culturally patterned (i.e., organized through social interactions) into recurring units which have been termed *literacy practices* by Scribner and Cole (1981), Heath (1983), Reder and Green (1983), and others. A cultural group may be said to have specific literacy practices in the same sense in which it is said to have specific religious practices, childrearing practices, and so forth. Scribner and Cole (1981) define a literacy practice as a

> recurrent, goal-directed sequence of activities using a particular technology and particular systems of knowledge . . . Literacy is not simply knowing how to read and write a particular script but applying this knowledge for specific purposes in specific contexts of use. (p. 236)

Examples of literacy practices include memorandum writing in an office, reading bedtime stories to children, liturgical recitation in a church, scanning the newspaper with morning coffee, review of a football team's play diagrams on a chalkboard, keeping personal diaries, and so on.

Several schemes have been formulated for identifying the constituent components of literacy practices. Scribner and Cole (1981) proposed three principal components: technology, knowledge, and skills. Szwed (1981) proposed five components of a literacy practice: text, context, function, participants, and (social) motivation. Reder and Green (1983) suggested three contextual dimensions of literacy practices: technological, functional, and social meaning. Although these (as well as other) classification schemes overlap considerably, Reder and Green's scheme is most closely related to Geertz's view of culture and the issues of cultural transmission being considered here. Their conception will be examined more closely below.

Literacy as Collaborative Practices

The analysis of literacy practices as collaborative activities is critical to the interactional perspective. Scribner and Cole's (1981) insightful analysis of literacy practices served their research purposes well—the analysis of the cognitive consequences of participation in particular literacy practices. Their analysis did not, however, systematically consider the dynamics of social interaction as a major dimension of context for literacy practices. Heath (1983), Reder and Green (1983), Shuman (1983), and Fingeret (1983) describe a rich variety of collaborative literacy practices. Individuals routinely share literacy skills and knowledge to accomplish their tasks much as they share carpentry, babysitting, and other skills and knowledge.

Reder (1987) distinguishes two senses in which literacy practices may be considered collaborative:

> A literacy practice may be collaborative in the sense that two or more interactants utilize written materials, as when members of a religious congregation read aloud as part of a responsive reading or family members alternatively read cards aloud from a board game being played. Alternatively, a collaborative literacy practice may involve a single participant who reads or writes in the performance of the task, but other members of the participant group nevertheless actively manipulate the information associated with that reading or writing. (p. 255)

Collaborative literacy practices occur within a variety of *participation structures*. In some collaborative literacy practices, for example, the participant group includes a *scribe* role; the specialized expertise, knowledge, or status of one participant combines with the writing skill, knowledge, or status of another (i.e., the scribe) as the group collectively accomplishes a task such as writing a letter or filling out a form. Szwed (1981) characterized another common role in literacy practices, the individual of unquestioned knowledge or authority, traditionally the *priest*. In still other practices, a different arrangement of specialization might occur: The person whose technical literacy skills are used varies from one occurrence of the practice to another. In yet other practices, the literacy skills of multiple participants are used simultaneously to accomplish the task at hand.

These various *social architectures* of participation cannot be understood simply as a byproduct of varying levels of individual skills and knowledge. Although in some cases the individual who acts as the scribe for the group may have better literacy skills than fellow participants, other factors may be involved, for example, participants' social statuses. Examples of collaborative literacy practice from domains of office work illustrate this point:

> consider the collaborative literacy practice of dictating a letter in a modern office setting. The 'boss' (one social role in this literacy practice) dictates the letter to the 'secretary' (another social role in the practice). The fact that the secretary rather than the boss is doing the actual writing (i.e., functions as the 'scribe') may or may not reflect the distribution of specialized literacy skills between the two; more likely it reflects their relative statuses within the organization in which they work. Similarly, when an individual asks an accountant or tax preparer for assistance in filling out income tax forms, it may or may not in a given instance reflect a specialization of knowledge or skills—it could merely represent a desire to spend one's time doing things other than filling out tax forms. (Reder, 1987, p. 256)

These examples of collaborative literacy practices, together with the many others described by Reder and Green (1983, 1985), Heath (1982b, 1983), Fingeret (1983), Shuman (1983), and Goody (1968), indicate quite clearly that a myriad of factors—not just individuals' level of proficiency at manipulating written materials—shape and are shaped by the participation structure of collaborative literacy practices.

Social Roles and Participation Structures

Ethnographers of communication traditionally focus on the role structure and interactive organization of participation in communicative activities in general, including literacy activities (e.g., Hymes 1968; Gumperz, 1982; McDermott, 1977). Participants' interactive roles and role-based expectations greatly influence the ways in which they conduct and accomplish their communicative activities. Differences between cultural groups have been powerfully drawn in terms of varying structure, roles, and modes of interaction within communicative activities (Scollon & Scollon, 1981; Philips, 1972; Heath, 1983). Differences between participation structures in home, school, and community settings have also been well documented (Heath, 1982a,b; Philips, 1972; Michaels, 1981; Shultz, Erickson, & Florio, 1982).

Within the realm of literacy practices, a similar approach holds promise of generating some important new insights. Heath (1983) examined the nature of participation structures in literacy activities in two working-class communities in Appalachia. Reder and Green (1983) studied the organization and workings of participant groups in a variety of literacy practices in an Eskimo fishing village in Alaska. Bloome and Green (1984) contrasted the social organization of literacy practices among urban junior high students in various settings. Taylor (1983) studied collaborative literacy practices in six suburban families and the ways in which (even among young children) literacy practices influence and are influenced by the system of social relations and statuses. These and other naturalistic studies of literacy practices exhibit a diverse set of participation structures across which a given individual may assume different roles and statuses.

Given the multitude of prevailing practices, contextual situations, and modes of interaction, a theoretical scheme is needed which can link characteristics of the social interactions used to accomplish a given practice with the ways in which participants' skills and knowledge are socialized to others. The concept of modes of engagement in literacy practices is useful in this regard.

Modes of Engagement in Literacy Practices

Drawing on comparative studies of literacy development across three ethnic communities, Reder (1987) suggested characterizing the social roles and interaction patterns underlying literacy practices according to individuals' modes of engagement in the practices. Reder suggested three distinct *modes of engagement*—technological, functional, and social:

> Individuals participate in these collaborative literacy practices in varied ways. Some persons directly manipulate written materials, reading and/or writing as part of performing the task at hand - these individuals are here said to be technologically engaged in the literacy practice. The term technologically engaged refers to the particular technology of writing involved in the practice. Other individuals may not be technologically engaged in the practice but nevertheless interact closely in performing the task with others who are technologically engaged—these individuals are said to be functionally engaged in the collaborative literacy practice. They may provide specialized knowledge and expertise as vital to the performance of the collaborative practice as the literacy skills of the technologically engaged participants. Other individuals may be neither technologically nor functionally engaged in the practice, but nevertheless have knowledge of the nature of the practice, and its implications for the life of the community, and must routinely take others' technological and functional engagement into account. Individuals in such positions are here said to be socially engaged in the practice. (p. 257)

A given individual may be engaged in the accomplishment of a literacy practice in various combinations of these three modes. Nearly all combinations are possible and occur frequently in many settings. For example, an individual may be technologically engaged but not functionally engaged—for example, someone performs the technological act of signing his or her name on a document without understanding the legal ramifications of so doing. Conversely, someone would be functionally but not technologically engaged in the practice of writing a letter when he or she dictates it to a secretary who writes it down.

Reder (1987) hypothesized that these three modes of engagement were closely related to distinct types of knowledge about literacy practices which Reder and Green (1983) had previously identified in the Eskimo fishing village: *technological knowledge, functional knowledge*, and *knowledge of social meanings*. Technological knowledge pertains to a particular technology of reading and writing used in literacy practices, and usually includes knowledge of the correspondence between units of the spoken and written language as well as knowledge of how to apply those correspondences in the context of the particular practice. Functional knowledge about a literacy practice refers to knowledge of how the technology is used for specific social purposes. Functional knowledge also encompasses the task and which individuals have the requisite technological skills for the task. The third system of knowledge, shared by all who are socially engaged in a literacy practice, concerns the social meanings associated with the literacy practice. This concept of social meaning requires some clarification.

Social Meanings of Literacy

In the analysis of communicative choice, sociolinguists such as Fishman (1972) have used the concept of social meaning to describe the social motivations underlying individuals preferences for using particular languages in given multilingual contexts. In a bilingual setting, individuals' use of, and alternation between, languages can be understood in terms of their social motivations to display greater or lesser affinities to social groups in which use of one language or the other is preferred. A language choice situation arises in communicative contexts in which interactants have multiple roles and statuses as members of social groups in which different languages are normally used or preferred. The choices made closely reflect participants definition of the situation, such as, which of their multiple statuses is most pertinent or appropriate. In collectively recognizing one set of their multiple statuses in this way, interactants dynamically establish and recreate the social meanings of language choice in the given situation.

There have been some recent attempts to extend and apply such this concept of social meaning to literacy (Stubbs, 1980; Spolsky, 1981; Szwed, 1981; Fishman, 1972; Reder & Green, 1983). In contexts in which multiple writing systems are in use—such as those in Spolsky's (1981) study of Navajo–English biliteracy on an Indian reservation in the Southwest, or in Reder and Green's (1983, 1985) studies of

Cyrillic–English, Hmong–English and Spanish–English biliteracies among ethnic minority communities—the social meanings associated with literacy choices may be understood in the same way. But the social meanings of literacy are also generated by communicative choices to use writing rather than other media of communication. In Reder and Green's (1983) study of literacy in the Eskimo village, for example, choices and alternations between use of writing and oral communication in various activities carried specific social meanings to the villagers.

Of course, great care needs to be taken in specifying relationships among particular social meanings and modes of communication. The relationships may be domain-specific and even practice-specific. Furthermore, sociolinguists have proposed cross-cutting distinctions among modes of interaction which may themselves carry distinct social meanings. Scollon and Scollon (1981), for example, have identified sharp differences in social meanings between "focused" and "non-focused" modes of group communication. Chafe (1982) and Iannen (1982) have proposed an "oral/literate continuum" of communicative styles, many of which naturally occur in both oral and written forms. The distinctive features of these varied styles (e.g., the degree of individual vs. group "focus" in Scollon and Scollon's work or in the degree of audience "involvement" in Chafe's or Tannen's account) are likely sources of social meanings generated by the collective accomplishment of the activities.

The sense of social meanings used here is closely related to Geertz's notion of culture reviewed above. These are the gossamer out of which the webs of meaning are produced through ongoing processes of social interaction; they are the stuff of which cultures are made. The next section considers how literacy is transmitted through these webs.

THE CULTURAL TRANSMISSION OF LITERACY

In this section, the processes through which literacy practices are socialized from one group or generation of individuals to another are considered. Three perspectives on socialization will be considered which are related to the cultural model of literacy practices developed in the preceding section: the *interactional* perspective, the *evolutionary* perspective, and the *ethnopedagogical* perspective. Each of these perspectives offers some interesting insights about the cultural transmission of literacy and some promising approaches to stimulating its development.

The Interactional Perspective

The interactional perspective emphasizes the importance of the participation structures of collaborative literacy practices. Participation in a particular social role or interactional mode within a given literacy practice is seen as the primary means of socializing the skills and knowledge associated with that practice.

Socialization of components. Reder (1987) suggests that the technological, functional, and social modes of engagement in a literacy practice are the contexts in which individuals acquire the corresponding technological, functional, and social meaning knowledge about the literacy practice. Socialization is thus dependent on individuals' access to those modes of engagement in the practice. This has a number of ramifications. One consequence is that inadequacies or inequities of access to critical participatory roles and modes of engagement are associated with corresponding deficiencies or inequalities of socialization and development.

Another implication is that knowledge of the technological, functional, and social meaning components of literacy may be socialized somewhat independently. Individuals may acquire one type of knowledge even though they have not developed another type. Functional knowledge of some literacy practices, for example, may develop well before the technological knowledge pertinent to those practices. This is true of a young child who has acquired much functional knowledge about story reading (e.g., books have pages to be turned, beginnings and endings, pictures and "stories," can be used to get a parent's attention) well before developing the technological encoding and decoding skills for reading those stories. A similar situation is that of a mechanic who has extensive knowledge about automobile repair before learning to read the technical books and diagrams others may use in making those repairs.

These situations contrast sharply with those in which technological knowledge pertaining to literacy practices develops ahead of functional knowledge. Western schools, for example, have been characterized by a pedagogy of literacy; the technological components of literacy have been taught in the academic context before students have been given the opportunity to use them in other contexts (i.e., before functional knowledge develops). Olson (1977), Cook-Gumperz and Gumperz (1981), Snow (1983), and others have suggested that this makes literacy development problematic for some students.

Literacy and social interaction. Researchers have described close links between the patterning of social interactions in which written materials are used and the development of literacy. Taylor (1983), Heath (1982b, 1983), Scollon and Scollon (1981), Michaels (1981), and Cochran-Smith (1984), among others, found close relationships between the ways in which adults and children interact with print and the development of literacy. The range of contexts and populations encompassed by these studies is impressive. Taylor studied suburban families; Heath studied two small, rural communities (one black, one white); Scollon and Scollon studied intercultural contexts and communication in Alaska; Michaels observed an ethnically mixed first-grade classroom; and Cochran-Smith studied a nursery school. The wide range of interactional routines and participatory modes reported is striking; no less significant, however, is the common finding that young children have early encounters with print in a wide variety of everyday activities in which they interact with parents, siblings, caretakers, and peers. The structure of these varied interactions in the school, home, and community contexts must be carefully examined to understand the ways in which skills and knowledge pertaining to literacy are being socialized.

Literacy and classroom interaction. Research about the interactional contexts of literacy events in the classroom has burgeoned in recent years. McDermott (1976) found distinct differences between the participation structure of high- and low-achieving reading groups in a first grade classroom. Low group members spent less time engaged in reading instruction than high group members, because of different turn-taking procedures; low group members' turn-taking procedures diverted attention and activities away from the instructional process.

Michaels (1981) also studied patterns of social interaction and discourse in a first-grade classroom. Her study focused on the analysis of one particular classroom routine, show-and-tell ("sharing time"). Students from different cultural groups displayed distinctive narrative styles during show-and-tell. Michaels characterizes this difference in terms of the degree to which the students' discourse was "topic-centered" (i.e., stays on one topic and verbally marks links between concepts). Michaels noted that the teacher responded differently to topic-centered narratives than to "topic-associated" narratives (i.e., multitopic narratives with implied or nonverbally marked relationships between concepts). As a result of this differential response to the students' narratives, they were, in effect, given differential access to "literacy learning events" in the classroom.

Michael's argument is that students bring different discursive styles from their home and community environments into the classroom. The teacher responds differentially to the display of topic-centered genres because those genres match the teacher's implicit model of what constitutes "literate" behavior or discourse. This selective reinforcement results in students having differential access to the social roles in the classroom through which they engage in literacy learning activities.

Wilkinson and Calculator (1982) found similar relationships between the language of classroom interaction and students' access to social resources for literacy learning. They looked at student interactions and discourse patterns within peer-directed reading groups, with a particular interest in the request and response interactions within the peer groups. Students who could structure requests in an appropriate way were more successful at obtaining needed information (and thus, Wilkinson & Calculator argued, access to literacy-learning resources).

DeStefano, Pepinsky, and Sanders (1982) also examined relationships between communicative behavior and reading instruction in an inner-city first-grade classroom. They looked closely at the language of reading instruction in the classroom. They observed that the teacher–student discourse about reading emphasized how students were to behave and how they were to become literate rather than providing direct assistance on the perception and comprehension of written materials. DeStefano et al. (1982) suggest that learning how to behave appropriate (from the teacher's perspective) in reading lessons rather than developing text comprehension skills may be the sina qua non of literacy in the first grade classroom.

Cultural differences. These studies, all based on careful observation of first grade classrooms, suggest that differences among students' modes of interaction and discourse in the classroom (with respect to those of the teacher as well as to each other's) may limit the access which some students have to literacy development resources (i.e., certain modes of engagement within classroom reading practices). During classroom reading instruction, students and teacher (like participants in all interactive situations) must define who they are and what they are doing. Based on these situational definitions, communicative rights and obligations are negotiated which structure participants' conduct and interpretation of the ensuing interactions. If participants' underlying definitions of situation and roles conflict, access to the literacy socialization process of the school may be severely limited.

There is serious reason to be concerned here, because numerous researchers have reported profound differences in the normative modes of discourse and interaction between cultural groups participating in the education system (e.g., Philips, 1972; Florio & Shultz, 1979; Scollon & Scollon, 1981; Heath, 1983). Differences these researchers have highlighted include discrepant situational definitions and attendant conflicts of social roles and communicative responsibilities between teacher and student as well as among students from different cultural backgrounds. The research on classroom interactions during reading instruction suggests that such differences may constitute a major source of difficulty in the development of literacy.

Conflicting social meanings. Several scholars have suggested that the primary problem facing young children learning to read in school is the transition from the "contextualized" language of the home and community (which is familiar to the child) to the "decontextualized" way in which written language is used in the school (Olson, 1977; Snow, 1983; Cook-Gumperz & Gumperz, 1981). From the interactional perspective, however, the problem may be not so much that language is being used without a referential or immediate experiential context as it is that the language that is being used has a conflictual if not highly negative *interactional context.*

According to the accounts of McDermott (1976) and DeStefano et al. (1982), in some classrooms the greater educational problem may not be that some students are lagging behind in acquiring the technological aspects of literacy, but that they are acquiring negative social meanings for literacy. Reder and Green's (1983) sociohistorical study of literacy development in an Eskimo fishing village in Alaska offers a powerful example of the intergenerational persistence of the negative social meanings attached to literacy. In their study, negative social meanings continued to inhibit literacy development long after the conditions disappeared that originally gave rise to those social meanings.

Such cultural conflicts may pose other obstacles to literacy development as well. Heath (1982a) and Gilmore (1983) offer powerful examples of how educators inadvertently but systematically misevaluate student literacy capabilities because of the ways in which the social organization of the classroom limits students' displays of competence at performing literacy tasks. Because of the limited repertoire of literacy practices acceptable in the classroom, students fail to display (and teachers thus fail to observe) competencies that are routinely displayed in informal peer games and activities in home and community settings. In Gilmore's (1983) powerful example of "Steps," a verbal game played by urban Black children, youngsters demonstrate communicative skills and

> language prowess, including abilities often unseen and unimagined in school contexts. (p. 253)

When students skillfully engage in literacy practices in contexts carrying positive social meanings but not in ones characterized by negative social meanings or uncomfortable participation structures, their literacy development is placed in double jeopardy. Not only are the negative social meanings present likely to be reinforced, teachers may misevaluate students' technological capabilities and thereby make inappropriate instructional decisions.

On the other hand, of course, it is possible for well-designed school reading programs, not only to avoid attaching negative social meanings to literacy, but to actually endow literacy with positive social meanings. If students have access to positively valued literacy-related interactions in schools, their literacy development should be facilitated. This is exactly what Au's (1980) research demonstrated. Au describes the social organization of reading lessons in an exemplary program at Kamehameha School in Honolulu, designed to meet the educational needs of Native Hawaiian children. The participation structures of the reading lessons were designed to harmonize rather than conflict with those the young children brought to the classroom from their home and community environments. When participation structures involving literacy are culturally appropriate, teachers can also avoid the systematic misevaluation of student capabilities described by Gilmore (1983) and Heath (1982a).

Evolutionary Perspective

The evolutionary perspective on the cultural transmission of literacy is concerned with how literacy practices naturally evolve or are stimulated to develop. The evolution of literacy practices, of course, may involve the use or incorporation of a technology of writing into an existing activity in which it previously has not been used (we shall term this *technology innovation)* or the innovation of a new activity altogether by using a technology of writing (termed here *practice innovation*). This section will be concerned primarily with literacy development as technology innovation, such as, the incorporation of written materials into existing cultural practices.

The newness of literacy development in a particular setting is a relative matter; the use of writing in the activity in question may be new to a particular individual, to a social group, or a society. As we move from one end of this continuum (the individual) to the other (the society), we move from individual literacy development to historical or societal literacy development.

Some historical examples. Let's begin by examining some historical cases of technology innovation in which a new communication technology was incorporated into existing social practices. The first case is an example taken from the history of the Vai tribe in West

Africa (Reder, 1977; Scribner & Cole 1981). The social practice in question is the sending of personal messages, which are usually sent between villages separated by a distance of at least several miles through the bush.

Prior to the introduction of their writing system in the early 19th century, Vai individuals "carried" oral messages between villages. Whenever one needed to send a message to a friend or relative in another village, one generally waited until someone was walking towards the village who could deliver the message. Ordinarily, the sender would recite the information that the messenger was to deliver; the messenger would travel to the intended recipient's village (sometimes en route to a more distant village), seek the recipient out, and recite the message. Sometimes, of course, the recipient was not in the village when the messenger arrived; in such cases, a third party would be asked to pass on the message, perhaps at the discretion of the messenger, perhaps according to instructions of the sender.

Depending on circumstances, other arrangements were made for message delivery. If urgency or confidentiality (or the lack of suitable travelers) so required, a runner would be sent specifically to deliver the message. If necessary, a chain of runners would convey the message over a longer distance, with each runner taking the message from one village to the next, reciting it to the next runner, until the message reached its ultimate destination/recipient. Each village chief in this communication network was responsible for conveying messages through his village's geographical territory.

The point here is that message sending was a well-established communicative practice, incorporated into the everyday life and social routines of the Vai, prior to the introduction of literacy. As the Vai developed an indigenous script during the first part of the 19th century (Koelle, 1854), one of its early functional uses was sending written messages between villages. Early specimens of written Vai messages reveal some fascinating information about their initial use. Rather than beginning with a salutation specialized to written correspondence (like the "Dear . . ." openings common to English language correspondence), early Vai correspondence (a few samples of which I was able to collect in 1976) featured salutations borrowed from oral message sending practices. For example, early letters began with a stylized salutation:

> Letter, you go from me to <recipient's name>. When you reach him, you tell him for me . . .

This written salutation is highly similar to the instructions the sender of an oral message would give to a messenger. The similarity between the two illustrates how a specific literacy practice, the sending of personal messages (still the primary functional use of the Vai script today), developed through incorporating the use of writing into an existing social practice. With time, these initially overlapping oral- and written-message sending practices began to diverge and specialize, such that there are now some types of messages that are usually delivered orally, others normally delivered in writing, and yet other types that are commonly delivered in either oral and written form. In concert with this partial functional specialization of the channels, written genres of Vai correspondence have gradually emerged, complete with their own characteristic salutations (a few elder Vai literates in the 1970s could still recall the older forms and salutations which written correspondence once took).

A second illustration of how a new communication technology becomes incorporated into existing social routines is taken from a more recent historical epoch. The setting is a large, bureaucratic government agency in the United States. The social practice of interest is the process of sending formal correspondence to a recipient outside of the agency. After a letter is initially drafted, it must be formally approved ("signed off on") by each successively higher level of the agency's organizational structure until it reaches the management level authorized to approve and sign the particular item. Different functional and topical categories of correspondence must be signed at varying levels of the organization. Considerable activity and concern are devoted to circulating materials to obtain the necessary approvals and signatures. Using the agency's intraoffice paper mail system, it can take many days for an item to be progressively routed and signed off on through the appropriate approval path in the organization. When items must be signed off on in a hurry, individuals often "walk them through", that is, hand carry them from one desk to another, getting the needed sequence of signatures.

Within the past 5 years or so, an office automation system has been installed and implemented throughout this agency. A new communications technology, electronic mail, has been introduced and is widely used. The system not only allows for the automated production, filing and delivery of written correspondence, it automatically routes documents from one person to another in the approval and signature chain of the organization.

Despite the savings of time and the increase of efficiency which this automated office system was expected (and already appears) to provide, users have incorporated it into existing communication practices much as the Vai incorporated the technology of writing into theirs. Despite the nearly instant electronic delivery of documents from one worker to another which the system provides, when a particularly time-critical document must be electronically "signed off on" and "out the door," individuals still "walk their document through" to expedite the process. In this case, however, it is an electronic document "walked through" rather than a written artifact that can be carried—one walks to the desk of the first person who must sign off on the electronic document, asks him or her to process the item in question (which will be waiting in his or her electronic inbox), then walks to the next electronic signatory's desk, and so forth. The use of a new communication technology has been incorporated into an existing social routine.

Despite the many profound cultural, historical, and technological differences between their settings, these two examples illustrate a similar evolutionary process for the incorporation of a new communication technology into existing social practices. Closer examination of each case reveals additional aspects of a general developmental pattern. As new technologies are first introduced, they initially tend to be used as alternative means of accomplishing existing, well-established social practices; this is a stage involving a high degree of *functional overlap* between the old and new techniques. As individuals become proficient with the new technology and familiar with its distinctive qualities and functional capabilities, experimentation and innovation tend to occur and new genres of communication unique to the medium begin to emerge. The range of activities in which the new technology is used progressively broadens until a functional equilibrium is established among the alternative ways of communicating (Reder & Conklin, 1987).

This evolutionary perspective offers fresh insight on the goal of stimulating literacy development. A promising strategy is to identify some existing cultural practices in which the use of written materials may be readily incorporated. This should give us an expanded sense of and concern with "promising practices." Literacy will more likely "take" (i.e., continue to grow and develop) when the use of the technology is initially incorporated into existing, well-established cultural practices, whose participants are permitted (if not encouraged) to experiment with and innovate new functional applications of writing. Strategies for accomplishing this are considered in Part 4.

The socialization of literacy among individuals. Similar principles may apply to the incorporation of literacy into the activities of the individual. Although many children learn to read in school through the interactive processes (and cultural constraints) previously described, many begin acquiring components of literacy through social interactions at home (Durkin, 1966; Teale, 1984; Heath, 1982b; Taylor, 1983), in preschool (Anderson, Teale, & Estrada, 1980; Wells, 1985; Cochran-Smith, 1984) and in other extraschool environments. Studies of emergent literacy in young children have identified such practices as parent–child storybook reading as the critical interactions in which the literacy behaviors are rooted developmentally (Teale, 1984, 1986). These studies of the fine microstructure of interaction between parent and child as they collaboratively read a story reveal a gradual shift of responsibility for accomplishing the literacy practice from parent to child as the child's proficiency develops. Vygotsky (1978) described such a transition as being from *inter*psychological functioning (the parent and child jointly construct the story) to *intra*psychological functioning (the child constructs/comprehends the story individually). It is clear from the careful ethnographic studies that Heath (1982b, 1986) and others have conducted on these emergent collaborative literacy practices that they develop in contexts carrying positive social meanings for the child.

Reder and Green (1985) argued that the socialization of literacy among adults occurs outside of formal instructional contexts in much the same way. Adults with poor literacy skills often collaborate with "helpers," trusted friends, neighbors, or relatives, to accomplish literacy tasks such as reading or writing letters, filling out forms, or reading interesting articles. Many of the collaborative literacy tasks described by Reder and Green (1983, 1985), Fingeret (1983), and Shuman (1983) share crucial properties with parent–child storytime reading. Reder and Green argued that, under appropriate circumstances, the helping relationships in which literacy tasks are jointly accomplished are learning contexts for adults. These concepts have strategic implications for literacy development that are considered below.

Ethnopedagogical Perspective

The *ethnopedagogical* practices of a cultural group are their ways of transmitting certain skills and knowledge (other than the given ethnopedagogical practice itself!). With regard to literacy, most societies and cultural groups have identifiable practices of this type. The "ABCs" or alphabet song is a familiar example of this in the English speaking world. Other cultural groups have also developed such songs, even in settings in which literacy is not taught primarily to children (e.g., the Vai of West Africa whom we considered above have developed several songs adults use to teach and learn their syllabary script).

There is not ample space to consider this topic in detail here. The important point is that there are numerous activities which serve an ethnopedagogical function without being so recognized by members of the society. Reder and Green (1985) describe a game played by young children in the Eskimo village they studied. Players in the game pretended they were sending or receiving letters from the village post office. Children scribbled make believe names and addresses of the sender and recipient on blank envelopes or folded paper serving as make believe envelopes. Children developed this functional knowledge of postal practices well before they mastered the technological skills which enabled them to actually produce intelligible written names and addresses.

Countless other games could be identified in this and other settings which serve to socialize various components of literacy practices. The next section will consider how these practices may be potentially useful contexts in which to stimulate informal literacy development.

IMPLICATIONS

The socialization framework and research considered in the preceding sections have sharp implications for how we should go about studying, as well as attempting to stimulate, the intergenerational transfer of literacy skills. Some of these implications pertain to literacy development in general, others to literacy training in schools, and others to informal literacy development outside of educational settings.

"Promising Practices": Context-Specific Literacy Development

The evolutionary perspective on literacy development considered in the previous section strongly suggests a process of literacy development that has yet to be exploited by instructional programs. If literacy development is viewed as the incorporation of the use of written materials into existing social practices, a variety of instructional strategies come to mind. Instructional approaches which attempt to facilitate the use of written materials in existing social practices have many advantages over approaches which seek to teach abstract literacy "skills" in one context with one set of materials but then leave it to students somehow to apply those skills in different contexts to different activities with different written materials.

This instructional strategy has been considered in detail by Reder and Green (1985). Among the theoretical advantages they see in this approach is its attempt to socialize literacy within familiar settings and interactional contexts, thus drawing on the positive social meanings transmitted in those contexts (as opposed to the negative ones sometimes carried by literacy in school contexts.) Another strength of this approach is its attempt to socialize the functional knowledge associated with a literacy practice together with the technological skills needed for its practical accomplishment. This again contrasts with the approach typical of many formal programs, in which technological skills and knowledge are taught without relevant functional content.

Optimal literacy instruction should be designed to draw on both the positive social meanings and functional knowledge inherent in existing social practices. Although literacy training approaches have been developed which combine technical and basic skills training for adults (Sticht, Armstrong, Hickey, & Caylor, 1987), there have been few systematic attempts to imbed literacy training in contexts carrying positive social meanings for otherwise hard-to-reach or hard-to-serve individuals.

An example of how such imbedding can take place was provided by Reder and Green:

> In a rural Black community a successful approach to teaching illiterates basic reading skills was based on putting materials (pictures of objects and their written names) on playing cards, which individuals commonly carry around and use in public within that community. By utilizing these socially acceptable materials (which do not outwardly mark individuals as illiterate), interest in learning to read was quickly elevated. (1985, p. 29)

Practice-specific literacy development should be cultivated in a range of activities and participant groups for maximum intergenerational impact. The example above, of course, involved introducing the use of written materials into the peer group activities of adults. The same strategy could be used to stimulate the use of written materials in cross-generational activities (e.g., a wide range of games played by parents and children). Identification of a wide range of "promising practices" such as these as potential contexts for literacy development should be a top priority for intergenerational approaches to literacy development.

Utilize Natural Helping Relationships

Reder and Green (1985) noted that most adults who do not have the requisite skills and knowledge to accomplish a literacy task seek out and cooperate with others who do have the needed skills and knowledge. These literacy "helpers" tend to be trusted friends, relatives or neighbors. They may help an individual to read a letter, fill out a form, scan newspaper ads, and so forth. These helping relationships—whether between adults or between adults and children—are potentially very productive contexts for imbedding literacy training. Reder and Green detail many reasons why such helping relationships and cooperative activities may be effective instructional settings, particularly for the hard-to-reach and hard-to-serve. "Helpers" could be the target for outreach efforts and then trained to build in their ongoing literacy assistance by gradually imbedding informal literacy training activities into their jointly accomplished tasks, much as some parents do while reading stories aloud with their young children.

Reder and Green dub the strategy of imbedding informal instruction into the naturally occurring settings, activities, and helping relationships of adults *giving literacy away*. It represents not only an alternative potential delivery system for the hard-to-reach and hard-to-serve, but a theoretically sound approach for stimulating the cultural transmission of literacy in general. The basic implementation pattern is:

> the process of drawing the curriculum of literacy training from the specific contexts of literacy helping networks. . . . Instead of developing curricula from arbitrary assumptions about what literacy tasks all adults 'must' be able to perform independently, begin with the particular literacy tasks and activities with which given adults are actively receiving assistance from natural helping networks. These social contexts are the promising precursors for literacy curricula, because they are already familiar to the illiterate participant, who already has supportive functional knowledge and social context for the particular tasks and materials. For example, if a common activity in a literacy helping network involves the use of shopping lists, then develop materials and teaching techniques around *those* shopping lists. If reading written messages from friends and neighbors is a task with which individuals naturally get help, then use *those* messages as the pedagogy and curriculum. (Reder & Green, 1985, p. 34)

Promote the Development of Positive Social Meanings for Literacy

Positive social meanings are the medium in which literacy develops. Instructional designers must take great care to insure that instructional contexts carry positive social meanings for the participants. Recalling McDermott's (1976) study of first grade reading groups, it is clear that literacy training does not merely succeed or fail at imparting socially abstract technological skills and knowledge; it simultaneously imparts varying social meanings to literacy. Too often the mark of educational failure is much more than the nonattainment of specific competencies, it is the attachment of negative social meanings to those competencies.

Conversely, Au's (1980) study of exemplary reading instruction for native Hawaiian children points out how the positive social meanings produced by culturally appropriate instruction can facilitate literacy development. The successes and failures of these varied programs should challenge us to discover the relevant systems of social meanings and culturally appropriate instructional formats for all of our student groups. As Gilmore (1983) eloquently points out, without this knowledge and insight, educators cannot productively assess the capabilities and educational needs of their students; but with that knowledge, doors may open to new and vital approaches which are relevant to and effective for their students:

> An examination of the sub-rosa literacy-related speech event steps provides dramatic evidence of these students' language prowess, including abilities often unseen and unimagined in school contexts. These events carry with them serious social and educational reminders. Successful and respectful pedagogies must be responsive to what children are capable of doing with language, both in terms of displaying competencies and in terms of conveying social messages. The implications are simple. The steppers have been urging us to listen for their plea to 'Gimme room.' They are asking for room to be seen as skilled language users and as individuals who have the right to instructional circumstances where pride and ownership are the central features of learning. In repeated chants, the girls advise all who are willing to hear that, given the respect and room they need, they are more than able to master the skills. (p. 253)

Incorporate Vital Features of Home/Community Participation Structures into Literacy Training Activities

One critical step towards endowing literacy with positive social meanings is to embed instruction in positively valued participation structures from the students' home and/or community. The work of Philips (1972), Scollon and Scollon (1981), Michaels (1981), Heath (1982a, 1983), Gilmore (1983), and others indicates the broad range of variation among cultural groups' normative modes of interaction, and how those differences can impede educational progress in the schools. It is certainly much easier to observe and acknowledge such cultural variability in their classrooms than it is to reorganize instructional activities in more effective ways in response to those differences (Mehan, 1980). But that difficulty must not impede the greatly needed experimentation, research, and development.

Attend Carefully to Multiply Structured "Messages" Carried by Literacy

The most extensively socialized components of literacy practices seem to be their social meanings (Reder, 1987). Indeed, there seem to be few (if any) historical instances in which literacy has been transmitted without a strongly associated system of social meanings. In the first-grade classroom studies reviewed above, children who appear to acquire few basic skills nevertheless learn clear social meanings for literacy. The contextual messages carried by these instructional practices are being clearly transmitted to all (not necessarily the same contextual messages, of course), even as the textual messages are not. The issue is not so much whether the message is getting across as what message is getting across.

Olson (1977) and many educational psychologists have argued that the major stumbling block to literacy development in schools is that the written messages valued by schools are inherently more "decontextualized" (with respect to children's everyday knowledge) than spoken messages. From the cultural perspective developed here, the problem may lie less with the quantitative degree of context-boundedness of textual materials as with the qualitative features of the social contexts in which they are used. Stated simply, we need to be concerned not only with how much context the reading materials and activities have, but also with what their interactional context is.

Recontextualize and Democratize Literacy Development

In summary, the cultural transmission of literacy takes place through multiple socialization processes in varied settings, contexts, and participation structures. Technological, functional, and social meaning components of specific practices typically are not transmitted as a unitary package; social meanings tend to be the most widely spread component. Although highly significant and valued in Western societies, school-based literacy practices are but one domain of literacy and but one of the processes through which literacy currently is, and likely will continue to be, culturally transmitted. In our pluralistic society, cultural differences attach varying social meanings to literacy; discrepant social meanings appear to be closely tied to inequities of access to literacy learning activities in school.

It is not prudent, however, to assume that schools in themselves can repair these powerful and deeply rooted conflicts of meaning, although in many contexts, more culturally appropriate instructional formats will facilitate literacy development. As we continue to look toward the schools for literacy instruction, a more comprehensive approach to the intergenerational transfer of literacy (and other cognitive) skills must be developed. We must broaden and democratize access to the array of contexts in which literacy development takes place. Materials, curricula, pedagogies, and delivery systems are needed to support formal and informal literacy development within a wide variety of social practices and participation structures. This recontextualizing of literacy must encompass activities in the schools, in communities, and in homes. Only then, when literacy is transmitted through diverse webs of interaction, social meanings, and cultural practices, will the rich messages of literacy get across.

REFERENCES

Anderson, A. B., Teale, W. H., & Estrada, E. (1980). Low income children's preschool literacy experiences: Some naturalistic observations. *The Quarterly Newsletter of the Laboratory of Comparative Human Cognition, 2*, 59-65.

Au, K. (1980). Participation structures in a reading lesson with children. *Anthropology & Education Quarterly, 11*, 91-115.

Bloome, D. (Ed.). (1987). *Literacy and schooling.* Norwood, NJ: Ablex Publishing Corp.

Bloome, D., & Green, J. (1984). Directions in the sociolinguistic study of reading. In P. Pearson, R. Barr, M. Kamil, & P. Mosenthal (Eds.), *Handbook on research on reading.* New York: Longman.

Chafe. W. L. (1982). Integration and involvement in speaking, writing and oral literature. In D. Tannen (Ed.), *Spoken and written language* (pp. 35-53). Norwood, NJ: Ablex Publishing Corp.

Cochran-Smith, M. (1984). *The making of a reader.* Norwood, NJ: Ablex Publishing Corp.

Cole, M., & Scribner, S. (1974). *Culture and thought: A psychological introduction.* New York: John Wiley and Sons.

Cook-Gumperz, J., & Gumperz, J. (1981). From oral to written culture: The transition to literacy. In M. F. Whiteman (Ed.), *Variation in writing: Functional and linguistic-cultural differences.* Hillsdale, NJ: Erlbaum.

de Castell, S., Luke, A., & Egan, K. (Eds.). (1986). *Literacy, society and schooling.* London: Cambridge University Press.

DeStefano, J., Pepinsky, H., & Sanders, T. (1982). Discourse rules for literacy learning in a first grade classroom. In L. C. Wilkinson (Ed.), *Communicating in the classroom.* New York: Academic Press.

Durkin, D. (1966). *Children who read early.* New York: Teachers College Press.

Eisenstein, E. L. (1979). *The printing press as an agent of change.* New York: Cambridge University Press.

Fingeret, A. (1983). Social networks, independence and adult illiterates. *Adult Education Quarterly, 33* (3), 133-146.

Fishman, J. A. (1972). Domains and the relationship between micro and macrosociolinguistics. In J. J. Gumperz & D. H. Hymes (Eds.), *Directions in sociolinguistics*. New York: Holt, Rinehart & Winston.

Florio, S., & Shultz, J. (1979). Social competence at home and school. *Theory into practice, 18*, 234-243.

Geertz, C. (1973). *The interpretation of cultures*. New York: Random House.

Gilmore, P. (1983). Spelling "Mississippi: Recontextualizing a literacy-related speech event. *Anthropology & Education Quarterly, 14*, 235-255.

Goodman, K. S. (1977). Acquiring literacy is natural: Who skilled Cock Robin? *Theory into Practice, 16*, 309-314.

Goody, J. (Ed.). (1968). *Literacy in traditional societies*. London: Cambridge University Press.

Goody, J., & Watt, I. (1963). The consequences of literacy. *Comparative Studies in Society and History, 5*, 27-68.

Gumperz, J. J. (1982). *Discourse strategies*. London: Cambridge University Press.

Havelock, E. A. (1963). *Preface to Plato*. Cambridge, MA: Harvard University Press.

Heath, S. B. (1982a). Questioning at home and in school: A comparative study. In G. Spindler (Ed.), *The ethnography of schooling*. New York: Holt, Rinehart & Winston.

Heath, S. B. (1982b). What no bedtime story means: Narrative skills at home and school. *Language in society, 11*, 49-76.

Heath, S. B. (1983). *Ways with words: Language, life and work in communities and classrooms*. Cambridge, UK: Cambridge University Press.

Hess, R., & Holloway, S. (1979). *The intergenerational transfer of literacy*. Washington, DC: National Institute of Education.

Hymes, D. H. (1968). The ethnography of speaking. In J. A. Fishman (Ed.), *Readings in the sociology of language*. The Hague: Mouton.

Koelle, S. W. (1854). *Outlines of a grammar of the Vej language*. London: Church Missionary House. (Reproduced Farnborough, UK: Gregg International Publishers, 1968).

Langer, J. A. (Ed.). (1987). *Language, literacy and culture*. Norwood, NJ: Ablex Publishing Corp.

Leach, E. P. (1982). *Social anthropology*. Oxford: Oxford University Press.

McDermott, R. P. (1976). *Kids make sense: An ethnographic account of the interactional management of success and failure in one first-grade classroom*. Unpublished doctoral dissertation, Stanford University.

McDermott, R. P. (1977). Social relations as contexts for learning in school. *Harvard Educational Review, 47*, 198-213.

McDermott, R. P., & Hood, L. (1982). Institutionalized psychology and the ethnography of schooling. In P. Gilmore & A. Glatthorn (Eds.), *Children in and out of school* (pp. 239-242). Washington, DC: Center for Applied Linguistics.

McLuhan, M. (1964). *Understanding media: The extensions of man*. Toronto: McGraw-Hill.

Mehan, H. (1980). *Learning lessons*. Cambridge, MA: Harvard University Press.

Michaels, S. (1981). "Sharing Time": Children's narrative styles and differential access to literacy. *Language in Society, 10*, 423-442.

Olson, D. R. (1977). From utterance to text: The bias of language in speech and writing. *Harvard Educational Review, 47*, 257-281.

Olson, D. R., Torrance, N., & Hildyard, A. (Eds.). (1985). *Literacy, language, and learning: The nature and consequences of reading and writing*: Cambridge, UK: Cambridge University Press.

Philips, S. U. (1972). Participation structures and communicative competence: Warm Springs children in community and classroom. In C Cazden, V. John, & D. Hymes (Eds.), *Functions of language in the classroom*. New York: Teachers College Press.

Reder, S. (1977). *The functional impact of writing on Vai speech*. Unpublished doctoral dissertation, The Rockefeller University, New York.

Reder, S. (1987). Comparative aspects of functional literacy development: Three ethnic American communities. In D. Wagner (Ed.), *The future of literacy in a changing world*. Oxford: Pergamon Press.

Reder, S., & Conklin, N. F. (1987, May). *The communicative economy of the workplace*. Paper presented at the 1987 Annual Meeting of the International Communication Association, Montreal, Canada.

Reder, S., & Green K. R. (1983). Contrasting patterns of literacy in an Alaska fishing village. *International Journal of the Sociology of Language, 42*, 9-39.

Reder, S., & Green, K. R. (1985). *Giving literacy away: An alternative strategy for increasing adult literacy development*. Portland, OR: Northwest Regional Educational Laboratory.

Schieffelin, B., & Gilmore, P. (Eds.). (1986). *The acquisition of literacy: Ethnographic perspectives* (Vol. 21). Norwood, NJ: Ablex Publishing Corp.

Scollon, R., & Scollon, S. B. K. (1981). *Narrative, literacy and face in interethnic communication*. Norwood, NJ: Ablex Publishing Corp.

Schultz, J., Florio, S., & Erickson, F. (1982). Where is the floor? Aspects of cultural organization of social relationships in communication at home and at school. In P. Gilmore & A. Glatthorn (Eds.), *Children in and out of school* (pp. 88-123). Washington, DC: Center for Applied Linguistics.

Scribner, S., & Cole, M. (1981). *The psychology of literacy*. Cambridge, MA: Harvard University Press.

Shuman, A. (1983). Collaborative literacy in an urban multiethnic neighborhood. *International Journal of the Sociology of Language, 42*, 69-81.

Snow, C. E. (1983). Literacy and language: Relationships during the preschool years. *Harvard Educational Review, 53*, 165-189.

Spolsky, B. (1981). Bilingualism and biliteracy. *The Canadian Modern Language Review, 37*, 475-485.

Sticht, T. G. (1983). *Literacy and human resources development at work: Investing in the education of adults to improve the educability of children*. Alexandria, VA: Human Resources Research Organization.

Sticht, T. G., Armstrong, W. B., Hickey, D. T., & Caylor, J. S. (1987). *Cast-off youth: Policy and training methods from the military experience*. New York: Praeger.

Stubbs, H. (1980). *Language and literacy: The sociolinguistics of reading and writing*. London: Routledge & Kegan Paul.

Szwed, J. (1981). The ethnography of literacy. In M. F. Whiteman (Ed.), *Variation in writing: Functional and linguistic-cultural differences*. Hillsdale, NJ: Erlbaum.

Tannen, D. (1982). The oral/literate continuum in discourse. In D. Tannen (Ed.), *Spoken and written language: Exploring orality and literacy* (pp. 1-16). Norwood, NJ: Ablex Publishing Corp.

Taylor, D. (1983). *Family literacy: Young children learn to read and write*. Exeter, NH: Heinemann Educational Books.

Teale, W. H. (1984). Reading to young children: Its significance for literacy development. In H. Coleman, A. Ober, & F. Smith (Eds.), *Awakening to literacy*. Exeter, NH: Heinemann Educational Books.

Teale, W. H. (1986). Home background and children's literacy development. In W. H. Teale & E. Sulzby (Eds.), *Emergent literacy: Writing and reading* (pp. 173-206) Norwood, NJ: Ablex Publishing Corp.

Vygotsky, L. S. (1978). The prehistory of written speech. In M. Cole, V. John-Steiner, S. Scribner, & E., Soubermann (Eds.), *Mind in society: The development of higher psychological processes*. Cambridge, MA: Harvard Univeristy Press.

Wagner, D. (Ed.). (1987). *The future of literacy in a changing world*. Oxford: Pergamon Press.

Wells, G. (1985). Preschool literacy-related activities and success in school. In D. R. Olson, N. Torrance, & A. Hildyard (Eds.), *Literacy, language, and learning: The nature and consequences of reading and writing*. Cambridge, UK: Cambridge University Press.

Wilkinson, L. C., & Calculator, S. (1982). Requests and responses in peer-directed reading groups. *American Educational Research Journal, 19*, 107-120.

Author Index

Subject Index